Finding the Balance
for earth's sake

Finding the Balance
for earth's sake

Dennis Minty   Heather Griffin   Dan Murphy

CANADIAN ENVIRONMENTAL SCIENCE SERIES
Newfoundland and Labrador Edition

BREAKWATER

Breakwater
100 Water Street
P.O. Box 2188
St. John's, NF
A1C 6E6

Front Cover Painting: Di Dabinett
Back Cover Photo: Dennis Minty

Canadian Cataloguing in Publication Data
  Minty, Dennis.
  Finding the balance
    (Canadian environmental science series)
    Includes index.
    ISBN 1-55081-065-0

  1. Ecology. 2. Environmental sciences. I. Griffin,
  Heather, 1959- II. Murphy, Dan, 1953- III. Title.
  IV. Series.
  QH541.M56 1993   304.2   C93-098587-7

Printed in Canada

Foreword

The main thrust of this book is the search for a reasonable approach in dealing with global ecological and environmental issues. Currently the care and protection of our environment are constant concerns. There are problems affecting our environment that just will not go away. We must first learn to understand these problems and then try to find ways to deal with them. There is no better place to begin than at home. Here is where we can begin to understand what is happening to our earth. Here is where we may find better ways to do things to protect our earth. Whatever we do in the way of understanding and action will be for earth's sake.

This book has been developed for the youth of Newfoundland and Labrador. It is the result of a vision held by a small group of people, comprising curriculum consultants, teachers, environmentalists and others, who feel that we, all of us, can do things better. In fact, there appears to be no choice. We must do things better. We must find the balance.

As publishers we have been challenged, as never before, by this awesome project. Our authors have given their best and we have been inspired by their sense of mission. We have made every effort to make the book itself a 'thing of beauty' that is in harmony with our environment. However, time and circumstances did not allow the publisher to pursue all the available paper products that would complete that harmony. There are other things that can be done. We are not yet satisfied. Perhaps as we take the message of this book out to young Canadians across the land we'll learn too and improve upon what we are offering our youth.

Meanwhile, enjoy this book. Learn from it. And do something for earth's sake!

Clyde Rose
Publisher

Newfoundland Author Team

Dennis Minty

Born in Twillingate, Newfoundland, Dennis Minty has a strong commitment to his home province. His post-secondary education at Memorial University of Newfoundland prepared him to be a wildlife biologist, but immediately after graduation he was drawn into the field of environmental education in which he has been a practitioner since 1973. His first professional work was to develop and manage a nature education centre, Salmonier Nature Park. He still remains as manager of the park, but is also Chief of Information and Education for the Newfoundland and Labrador Wildlife Division. Also a published photographer, Dennis lives with his family in St. John's.

Heather Griffin

Heather Griffin received her undergraduate degree in English and Philosophy, followed by a second degree in Outdoor and Experiential Education from Queen's University. Heather's philosophy in teaching and curriculum development is that it is important to blend the arts, particularly music, with environmental education. She is a talented musician who has had work recorded and has also been featured in a number of radio and television productions. Heather is currently Program Director for Newfoundland and Labrador for the Atlantic Centre for the Environment. She lives with her husband and son in St. John's.

Dan Murphy

Dan Murphy was born in Noranda, Quebec, grew up in Northern Ontario, and came to Newfoundland in 1972 to study biology at Memorial University. He holds a B.Sc., B.Ed. and a Diploma in Industrial Arts Education–all from Memorial University of Newfoundland. He has recently completed a Masters of Science Education with a major in Outdoor Teacher Education from Northern Illinois University. He has been teaching science at the junior and senior high school levels for the past sixteen years. Dan has also taught computer and photography courses for the past ten years with Sir Wilfred Grenfell College Extension services. His major area of interest is outdoor education.

Acknowledgements

This book and the Teacher's Resource that accompanies it are the results of many people's dedicated efforts. The authors would like to thank them and acknowledge their contributions.

As this work was researched, compiled, written, edited, designed, illustrated, pilot tested, reviewed, rewritten, and re-edited, all in the span of thirteen months, there was little time left to share with our families. We therefore thank them for we are truly grateful for their tolerance and support during this hectic period.

We also want to acknowledge the outstanding work of the staff at Breakwater Books and the commitment of its president, Clyde Rose, to the project. Celie Giovannini, the production manager, showed uncompromising commitment to excellence in fulfilling her responsibility for design and layout. Patti Giovannini not only served as a sensitive and competent editor, but also contributed in handling the hundreds of details associated with a project of this scope. Chris Charland, Celie's main help, diligently tackled the computer entry and physical layout, and Mona Abbott-Kesting meticulously proofread the entire work.

Certain staff of the Department of Education deserve special mention for their roles. Dr. Harry Elliott, Barry LeDrew, Jacinta Sheppard, Wilbert Boone and Dr. Wayne Oakley should take pride in their influence on the final product. Within their collective minds, the idea for this project arose. They showed foresight, courage, and a progressive attitude towards education about the environment, all of which contributed to the making of this book.

We also acknowledge the members of the Environmental Science Working Group—Dr. Tony Blouin, Ms. Pamela Constantine, Capt. Philip McCarter, Mr. Claude Schryburt, Ms. Pamela Walsh and Mr. Don Wight—and the pilot teachers who helped with the fine tuning of the text and teacher's resource book; Michael Holmes for his tireless researching and compiling information on each of the issues; Bill Montivecchi, Jon Lien, Susan Ahearn, and Bill Meades for reviewing various portions of the work; the many people who provided technical reviews of specific chapters; Sue Meades for her precise technical illustrations, contributions to the Teacher's Resource, and editorial comments; Sylvia Ficken for the more light-hearted art work; Di Dabinett for her captivating cover art; and the Atlantic Centre for the Environment for its ongoing support and commitment to the project.

This book, among the first of its kind in Canada, also represents the first science text to be published in this province, about this province and for this province. All those involved should be proud.

Dennis Minty, Heather Griffin and Dan Murphy

Table of Contents

UNIT 1
Basic Ecological Concepts

THE ECOSYSTEMS OF NEWFOUNDLAND AND LABRADOR

Unit 1

This first unit of the course looks at the nature of environmental science as well as many of the important ecological concepts underlying the environmental problems facing our planet today. It should provide you with a basis for understanding the issues that lie ahead in the following units, and those you encounter from day to day in the popular media.

These first two chapters begin with a look at the earth from outer space—viewing the earth first as a whole, then focusing inward towards the surface of the planet. Finally, we look at specific ecosystems and regions of the globe.

We hope you enjoy this view of your home planet. May the forces of nature be with you!

What you will learn:

- an understanding of the nature of environmental science;
- a definition of the term 'sustainable development';
- an understanding of the concepts of biosphere, terrestrial biomes, and aquatic realms;
- some basic concepts of the science of ecology, including ecosystems, communities, interdependence, the cycles of nature, energy flow and population dynamics;
- the ecosystems of Newfoundland and Labrador, with a close look at wetlands;
- the idea of ecoregions, with a specific look at the ecoregions of this province.

CHAPTER 1
Introduction To Environmental Science

Unless. Unless someone like you cares a whole awful lot, nothing is going to get better. It's not.

Dr. Seuss
The Lorax

The Nature of Environmental Science

What You Will Learn:

- the nature of environmental science—what it is, what it isn't, and why it's important to all of us;
- the earth as a spaceship—the concept that the earth provides us with all the basic elements needed for life;
- a relatively new theory—the Gaia theory—that presents a different look at the functioning of the planet;
- the ideas of biosphere, biomes, and ecosystems;
- the concept of interdependence—that all things in life are connected to and dependent upon other things;
- an introduction to the term 'sustainable development'—attempting to balance the needs of humans with the needs of nature;
- the earth's great cycles—water, nutrients, and others;
- the flow of energy on the planet—including a look at the feeding connections between plants and animals.

The future of our planet and all life on it is threatened by the relatively recent actions of people. Practically every day we hear news that the quality of our air, water and land is worse than it was a year ago. Although there are some improvements, they are slower than the rate of deterioration. Yet the earth is the only home we have! We must learn to live on it wisely and act in ways that support the environment.

Sustainable actions and projects are those which find a balance between serving our needs and keeping the earth healthy. The earth can support these actions without too much damage. Other actions and developments, such as the production of harmful chemicals that damage the atmosphere, are unsustainable. You will hear these two words 'sustainable' and 'unsustainable' quite frequently in your study of environmental science, so understanding their meaning is essential.

We study environmental science to learn how to make responsible decisions about our own future and that of our children.

Activity: You and the Environment

You, as an individual and as a citizen of the world, have already formed many ideas and opinions about life on this planet. Some of these may change as you encounter other people and experiences, perhaps even as you complete this course.

Throughout this year you will be keeping an environmental journal—a separate notebook in which you can record your thoughts, observations, sketches, and other things. Try to bring to the surface some of the thoughts and feelings you may already have about the environment. On the first page of your journal, write down any ideas that come to mind. Express yourself through writing, drawing, poetry...loosen up your mind! What does the word 'environment' mean to you? What feelings do you have when you're alone with nature? Do you see yourself as part of the natural world, or separate from it? How does it make you feel when you hear about the environment in the news? Remember, this is not a test. It is simply a chance to reflect on you and the environment.

Think About It:

We do not inherit the earth from our grandparents, we borrow it from our grandchildren.

Lester Russell Brown

What does this mean to you?

Environmental science is still quite new. It has grown out of another relatively new science—*ecology*—which is the study of the interrelationships among living things and their environment. Environmental science builds on ecology by adding social, ethical, political, legal and economic factors to the study of natural systems. It is primarily concerned with the effect of people's activities on the earth—its plants and animals, its rivers and oceans, its forests and barrens, the air we breathe, and many other elements of life on our planet. Environmental science refers to the process of applying ecological principles—the natural workings of the planet—to human use of the environment.

Unlike physics and chemistry, environmental science is not made up of a clear set of rules or facts. Although there is much we know about our environment, there is far more we don't know. Therefore, the topics we study in environmental science are often controversial and involve different points of view. In most of these situations there are no simple solutions, no clear right and wrong answers. The solutions usually involve compromise. They also require careful thought and prediction of results based on the best information available, even when there is so much unknown. These are some of the reasons why studying environmental science is so interesting.

Environmental science is not separate from other sciences like biology, ecology, physics, chemistry, mathematics, and earth science. Rather, it combines all of these disciplines with social, political, legal and economic concerns. It also takes into account your values and attitudes—what matters to you and how you think about it. Just about everyone will say that a healthy environment is important, but what are you prepared to do about keeping it healthy? What do you do that hurts the environment? How are you prepared to change your actions to show that you value your environment?

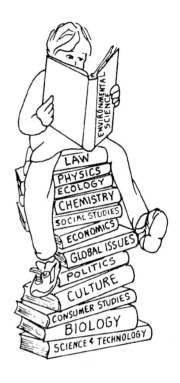

Just about everyone will say that a healthy environment is important, but what are you prepared to do about keeping it healthy? What do you do that hurts the environment? How are you prepared to change your actions to show that you value your environment?

17

Ecological processes: the relationships among living organisms and with their nonliving environment, including energy flow, and water, gas and mineral cycles.

Because environmental science covers such a broad base of knowledge, methods and values, you may feel intimidated by it at first. But it is important for you to stick with it. Environmental science is one of the most important areas of study to help you become a full, responsible citizen, no matter what your career choice. It is equally important for an engineer, teacher, farmer, doctor or bulldozer driver to understand something about environmental science. Much of what you will learn in this course concerns your everyday actions and choices.

Many of our environmental problems, like acid rain and global warming, can appear immense and make ordinary people feel powerless to fix them. However, the first step is to understand the problems. That's why we study environmental science.

No one person can make all the necessary changes to ensure the health of the earth. But if each of us alters our behaviour in a few small ways, these changes can add up to a significant improvement.

Sustainable development generally means the development of our resources to meet our present needs without reducing the ability of future generations to meet theirs. It requires that we live within our means. By placing limits on what we do and how much we take from the environment, we can ensure that basic **ecological processes** are not hurt. Sustainable development also links the health of our environment with the health of our economy. One becomes impossible without the other. Sustainable development is now widely accepted as the correct course for all societies to follow.

Activity: Building Our Environmental Awareness

We hear about the environment all the time—in the news, in advertising and from our politicians. Some of it is good, some bad. Set aside a wall at the back of your classroom, or a class scrapbook, or even a section of your own Environmental Journal. Make this your Environmental Awareness area. Bring in articles, photographs or pictures. Keep building on this area as you complete this course!

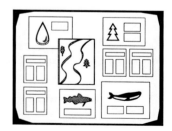

Sustainable development is a relatively new idea that has taken us many generations to come to. Its increasing importance to today's society is largely due to the world's continually expanding population, which has placed increased demand on our natural resources and our available natural spaces. It is up to the present and future generations to make it work. This is why the study of environmental science must include a focus on sustainable development.

In this course, you will be looking at many environmental problems, but don't feel overpowered by their number or size. By applying the principles of sustainable development, we can all help solve these problems.

The search for a new balance between planetary ecology and human development is more than just a search for new technologies and new scientific theories. It is an attempt to reconcile different ideas about what goals we should pursue.

A Primer on Environmental Citizenship, Environment Canada, 1993

Analysis:

1. What is environmental science?

2. You have already completed some science courses. What aspects of these courses do you think might apply to environmental science?

3. What are some human behaviours that contribute to the state of the environment?

4. What is sustainable development?

5. Name two natural resources in Newfoundland and Labrador to which the concept of sustainable development can be applied. Is it being applied now?

There's so much that we share, that it's time we're aware—It's a small world after all.

It's a Small World

19

When man went into space, he looked back and discovered earth.

"Nova," PBS
Science Program, 1991

Our Earth

Imagine all the grains of sand on a huge beach. Think of each grain of sand as a star or planet in the universe. Now imagine that life exists on just one of those grains of sand. As far as we now know, the combination of factors that allow life to exist are present on only one small planet out of a seemingly infinite number of celestial bodies. That planet is Earth.

The first pictures of earth taken from space caused people to rethink their place in the world. It was like a flea on a dog's back getting a view of the whole dog for the first time! Until then, the flea's entire world was made up of the warm skin of the dog and the shelter provided by the thick fur. The flea would not have known where the dog was or what it was doing most of the time.

Like the flea, people are often aware of their immediate environment but have difficulty thinking of the earth as a whole — of how all the parts are connected, and how it works.

The pictures from space allowed us to see our world as a beautiful multi-coloured globe surrounded by black space. Now we are beginning to realize that it is quite small, that it has real limits and that it is unique in the universe. Just like a space capsule carrying the astronauts, the earth is a self-contained unit in itself.

Astronauts depend on the limited resources they carry aboard their space craft. They take on board all the water, food and oxygen they need for their flight. If they run out or contaminate what they have, they cannot replace it from a new source.

The earth's resources are very similar to those of the space craft. The earth is a **closed system**, like a spaceship, since there is no input of new resources, except solar energy. Like the astronaut, we must conserve and care for these resources, or our space craft, the earth, will fail.

Remember, if conditions on earth become unbearable, there is nowhere else to go. The environments of all other known planets are not suitable for human habitation. The possibility of maintaining large human populations elsewhere in the universe is not as real as it may appear on "Star Trek".

Aboard the space craft, as well as on the earth, it is important to maintain a state of **homeostasis**. Homeostasis means dynamic balance or, in other words, a stable, sustainable condition while many things are changing

Closed system: a system in which there is no, or minimal, input of new material from the outside.

Homeostasis: the maintenance of a stable condition that depends on many interactions in order to remain stable.

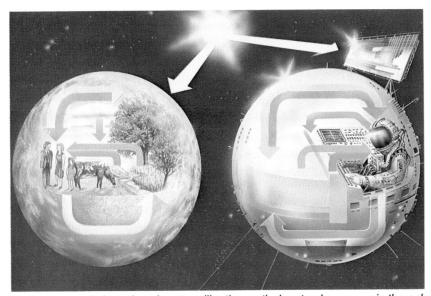

A space capsule is a closed system like the earth. Input solar energy is the only new resource. All gases, water and other elements that are necessary to support life are constantly recycled.

Ecosystem: a self-supporting community of plants and animals interacting with each other and the nonliving environment to produce a balanced system. Since it is more an idea than a place, an ecosystem can be as small as a pond or as large as the earth.

NASA: the National Aeronautics and Space Administration, which is responsible for many American space flights.

both inside and outside of it. Consider your own body. There are constant changes occurring as you take in oxygen and expel carbon dioxide. You are digesting food as blood pumps through your body, feeding your brain and other organs. You are seeing, feeling, acting or reacting all of the time. Yet, overall, your condition is stable. In fact, it would become very unstable if, for some reason, one of the important changes taking place in your body failed to occur. If your liver or kidneys stopped working properly to help your body deal with waste products, you would die unless it was corrected promptly. Just like your body, natural systems require constant change and must work together to remain stable. This is homeostasis.

When you look for homeostasis in the earth **ecosystem**, it is difficult to see. Individuals do not live long enough to detect any disturbance in global stability —it happens over several generations. Yet we have learned recently that the earth's homeostasis is threatened. If you were looking at the control panel of the earth space ship, its warning lights would be flashing!

The goal of sustainable development is to ensure that homeostasis is achieved for the earth. Most ecosystems are strong enough that some processes can be damaged and some components can be lost without the whole system failing. But we don't know exactly which components we can afford to lose, and we cannot be sure how much of the earth can be hurt before the results are disastrous. Once systems break down it is very difficult and very expensive to repair them. It makes far more sense to keep them working well in the first place.

The Gaia Theory
The Earth is Alive!

Dr. James Lovelock, a British inventor and scientist, proposed the Gaia theory in the 1970s, naming it after the Greek goddess of mother earth. While working for **NASA** in the late 1960s, he used his invention, the telebioscope, to examine the atmosphere of Mars to determine if it could support life. He concluded that it could not. He then examined the earth's atmosphere and discovered that it was composed of a mixture of gases created by the plants and animals living on the earth's surface. He realized that the life processes of these plants and animals *are required* to keep the atmosphere in a state that will support their existence. He used this observation to propose the Gaia theory.

The Gaia theory states that the earth functions like a living organism, not simply an environment where life exists. It is a self-sustaining system that modifies its surroundings to maintain the climate, atmosphere, soil and water in a balance that is favourable to life.

Lovelock argued that living organisms are influenced by and adapted to their environment, but they also change their environment to make it more hospitable on a global scale.

Many global characteristics support this theory. The amount of oxygen in the atmosphere has been stable for 200 million years. The surface temperature of the earth has been relatively constant for four billion years, even though the sun's heat has increased 25% during this time.

The traditional view of evolution, based on natural selection, suggests that as conditions on the planet change, life either adapts to it or dies. In the Gaia theory, this might be restated as: life either fits into the functioning of the whole earth and contributes to it or it will be eliminated. Species which don't support the proper working of the earth will not survive.

Even though the Gaia theory is still controversial, it does provide a useful perspective to help develop an understanding of the earth.

...the complexity of the Earth and its life systems can never be safely managed. The ecology of the top inch of topsoil is still largely unknown, as is its relationship to the larger systems of the biosphere. What might be managed is us: human desires, economies, politics, and communities...It makes far better sense to reshape ourselves to fit a finite planet than to attempt to reshape the planet to fit our infinite wants.

David Orr

Earth stewardship: the act of looking after the earth as you would your home.

Think About It:

1. *Do you think the earth is fragile? Explain your answer.*

2. *Research the Gaia theory further and describe Lovelock's Daisy World.*

3. *Many world religions promote the Gaia theory since it supports the idea of **earth stewardship**. Explain your view of the possible connection between a religion and the Gaia theory.*

4. *Research the spiritual orientation towards nature of one or more of Canada's indigenous peoples. Assess your findings in relation to the Gaia theory.*

The Biosphere

Carbohydrate: an energy-rich substance made from carbon, oxygen and hydrogen—usually a starch or a sugar.

As far as we know, in all the universe, life exists only on earth. Even here, life thrives only within an amazingly thin skin of the earth's surface and the air above it. This tiny, unique fraction of the universe is the *biosphere*. It consists of three layers: the *atmosphere* (air), the *hydrosphere* (water) and the *lithosphere* (rocks and soil). Life is found in and on these layers. The biosphere is only about 15 kilometres deep—one four-hundredth of the earth's radius.

The life-supporting biosphere is warmed by the sun to a relatively narrow temperature range—between about -40°C and +40°C. Although this may seem pretty extreme, consider that the planet Pluto can be as cold as -200°C and the planet Mercury can be as hot as +480°C. These are planets that share our sun! Clearly, the biosphere is unique, and its ability to sustain life is very precious.

Almost all life on earth depends on energy from the sun. Green plants are the 'food factories' that use light to convert carbon dioxide and water into simple sugars. These sugars, in combination with other nutrients, are converted into protein, oils, and **carbohydrates**. Animals then absorb these nutrients by eating plants.

24

Plants and animals, including humans, are the living parts of the biosphere. They interact with each other and the nonliving parts of the biosphere, like sunlight, soil minerals, bodies of water and gases that make up the air. The living parts of the biosphere are known as *biotic factors*, while the nonliving parts are *abiotic factors*.

When plants and animals die and decompose, they release gases and nutrients that are reused by new plants and animals. This is all part of the cycle of life—the reuse of tiny portions of the earth's storehouse of materials and the transfer of energy. Recycling was not invented by people—the earth has been doing it for millions of years!

People are one species among millions that share the thin veil of life surrounding the earth. We are one of the biotic factors in the biosphere. We depend on the health of the biosphere just like the lions of the African savannah, the penguins of Antarctica, the fish in our oceans and the earthworms in our gardens.

Yet, for an 'intelligent' species, it is shocking what we are doing to the biosphere. We release toxic wastes into the water and the skies. We destroy millions of hectares of tropical rain forest. We burn fuel at such a rate that the carbon dioxide build-up in the atmosphere exceeds the rate at which it can be removed. We are causing plants and animals to become extinct faster than ever before in the history of the earth.

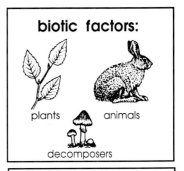

biotic factors:

plants   animals

decomposers

abiotic factors:

sunlight   weather

soil and water

Activity: An Apple and the Earth

Consider an apple as representing the planet earth. It must meet the needs of all living things by providing food, water and shelter. What portion of this earth provides habitat for humans and other land-dwelling animals?

- About 75% of the earth is covered by oceans, so cut away three quarters of the apple and keep the remainder.
- Thirty percent of the remaining quarter is desert, unfit for human living, so cut off a third and put it aside.
- Another 30% is mountains, too high to live on, so cut off another third and put it aside.
- Of this remaining portion, only the surface (skin) is used as living space for humans and many land-dwelling animals, so trim the skin from a small piece of apple. This piece of skin represents the amount of the earth's surface that we have available to us as human habitat. Even this we do not have exclusive rights to, since we share it with many other creatures.

How long will it be before our population consumes all of this available living space?

Decomposer: bacteria, plants and animals that feed on dead plant and animal tissue.

E.O. Wilson, a famous biologist, once found over 200 species of ants in a single tree of a tropical rain forest. Only about one-half of these species had been previously recognized by scientists. This is an example of how incomplete our knowledge of the biosphere really is. Many species of plants and animals are disappearing before they are even studied and identified, and before their role in the biosphere is understood.

Later in this book we will cover these topics in more detail. For now, we must recognize that humankind is part of the biosphere, not separate from it. We are completely dependent on a healthy biosphere, yet we are the one species that poses the greatest threat to it.

Activity: A Piece of the Biosphere - Up Close

You now know that the biosphere refers to the surface of the earth, its water, and the air above it. Take a bit of time to examine a very, very small piece of that biosphere, somewhere near your school or home.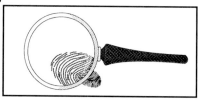

Use some string to outline a square piece of ground, one metre by one metre. Use a magnifying lens to carefully examine each part of this ground. In your journal, record any insects you find, any rocks, and the variety of plant life you encounter. What else do you see or feel? Close your eyes and listen. What do you hear? Keep your eyes closed and feel the earth around you and under you. This is your earth—one that you share with all the other things you discovered in your small piece of the biosphere.

Analysis:

6. What is the biosphere? Name its three parts.

7. List three local plants and animals that inhabit each part of the biosphere.

8. Using knowledge you have picked up from the newspaper, radio, and television, briefly describe three problems facing the biosphere.

9. What are three biotic factors and three abiotic factors that influence your community?

A Closer Look: Biosphere 2

In 1991, at a location near Tucson, Arizona, four men and four women (one of whom was a former student at Memorial University, Abigail Alling) entered an air-tight, glass and steel structure covering an area a little larger than two football fields. The door was closed and sealed, and this enclosure became their world for the next two years.

No, this was not a prison. The men and women were scientists—the select crew of Biosphere 2, a prototype space station undergoing its testing here on earth. Their mission had two purposes. One was to test systems for setting up permanent human communities on other planets. The other was to collect information for managing 'Biosphere I', our own spaceship Earth.

Biosphere 2 was an ecological experiment. No food, water, oxygen, or other life-support systems entered the enclosure after it was sealed. Instead, the 'Biospherians' attempted to live by the same principles that sustain the earth.

Their enclosure contained 4,000 species of plants, small mammals, birds, reptiles, insects and soil microbes. These species were arranged to create samples of tropical rainforest, savannah, desert, scrub forest, fresh and saltwater marshes, and a mini-ocean complete with a living coral reef.

Sunlight sent currents of warm air across the ocean, causing evaporation. The moisture would condense and create high rainfall over the tropical rainforest. The water would then trickle back towards the marshes and ocean, through soil filters. This created a continuous supply of purified fresh water. The biospherians grew vegetables to feed themselves as well as a few goats and chickens, which in turn produced milk and eggs. Aquaculture (fish farming) supplied additional protein.

With everything sealed in a relatively small space, you might wonder how all the human and animal wastes would be disposed. This too was all recycled so the nutrients could go back to supporting the growth of plants.

Sound like a fun way to spend a couple of years? Probably not for all of us, but some of the lessons learned from this experiment may provide valuable information about how the cycles of the earth actually work.

The task of duplicating nature, as in Biosphere 2, is a tremendously difficult and expensive one. Spaceship Earth is our own precious and highly complex life-supporting system. Doesn't it make sense to ensure its stability? Unless you particularly like the idea of living in a bubble, that is.

Biomes and Ecosystems

It is difficult to study the biosphere as a single unit. To understand the many factors influencing the biosphere, scientists have broken it down into parts called *biomes*.

Terrestrial (land) biomes are large portions of the earth with similar climate, soil, plant and animal communities (see Figure 1.1). Where climate and landscape are similar, the same communities develop. These large communities are what we call biomes.

There are two water biomes or *aquatic realms*: oceans and fresh water. Each of these realms is usually considered in still smaller portions. For example, fresh water can refer to rivers, streams, lakes and ponds. Similarly, we can subdivide oceans into the coastline, continental shelves, the ocean surface, the mid-water depths, and the very deep ocean bottoms.

A Closer Look: building biomes

In 1992, there was an interesting project underway in Canada. The city of Montreal was developing plans for the duplication of four major world ecosystems—under one roof. The location for the project was right next to the Olympic stadium, in what was once the Olympic bicycle track—the Velodrome.

The project was called the Biodome of Montreal. The goal was to duplicate nature as closely as possible—essentially squeeze portions of the world into a confined space.

Visitors would be able to see and learn about the tropical forest, a northern forest, a marine ecosystem, and the polar world. There would be representative species of plants and animals, the climate conditions would be simulated, and bodies of water (fresh or salt) would be included.

Try to imagine that you were a designer for part of one of these projects. Let's say you had to build the tropical rainforest

right beside the polar ecosystem. What kinds of challenges would you face? Try to sketch a design that would take into consideration not only the needs of the animals and plants living there, but also the needs of visitors to view and learn about these places.

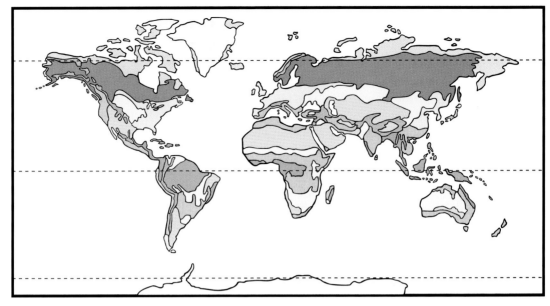

Figure 1.1 - Biomes of the World

- ▨ Tropical Rainforest
- ▨ Tropical Deciduous Forest and Temperate Evergreen Forest
- ☐ Temperate Deciduous Forest and Mixed Deciduous/Conifer Forest
- ▨ Boreal Forest
- ☐ Grasslands
- ▨ Desert and Semi-desert
- ▨ Mountains
- ▨ Tundra
- ☐ Polar Ice

Desert

Ocean

Freshwater

Forest

Grassland

Tundra

Interdependence: the relationship that animals and plants have with each other and other elements of the environment.

Food Web: an interlocking pattern showing the eaters and the eaten. Each member of a food web is classified as one of the following:

- **producers** - green plants;
- **primary consumers** or **herbivores** - animals that live mostly on green plants;
- **secondary consumers** or **carnivores** - animals that eat herbivores;
- **tertiary consumers** (also carnivores) - animals that eat other carnivores;
- **omnivores** - animals that eat both plants and other animals;
- **decomposers** - bacteria, plants and animals that feed on dead plant or animal tissue.

Energy chain: the transfer of energy from one level of an ecosystem on to the next.

Ecotone

Both aquatic realms are well represented in Newfoundland and Labrador. The terrestrial biomes found here are the tundra and boreal forests. Tundra exists mainly at the tops of some of our mountains and in northern Labrador. The boreal forest biome occurs throughout central and southern Labrador and most of the island. We will study the natural communities of this province in Chapter 2.

The Nature of Ecosystems

Everything is connected to everything else.

Barry Commoner

In all ecosystems, plants and animals depend on other plants and animals and their nonliving environment. They live in a web of **interdependence** in which each species contributes to and relies upon the working of the whole system.

The relationships or connections in an ecosystem are best understood by examining **food webs** and **energy chains**. Food webs illustrate who eats whom in an ecosystem. For example, Figure 1.2 on page 31 shows a simple food web in a forest.

The boundaries between ecosystems are fuzzy. There is no wall between a forest and a bog. Instead, as you move from a forest onto a bog, the trees get shorter and bog plants become more common as a result of changing drainage and soils. We call these areas between different ecosystems *ecotones*. Ecotones have characteristics of both ecosystems to which they are connected. For example, they have plants and animals that might be from either the forest or the bog.

Ecosystem means more than just living and nonliving things. There are also the connections between all these things. The flow of energy, the cycling of nutrients and water, the eaters and the eaten, death and life — these are a few of the relationships at work in an ecosystem. We will look at these ideas more closely to begin to understand the delicate and complex way in which all things are connected.

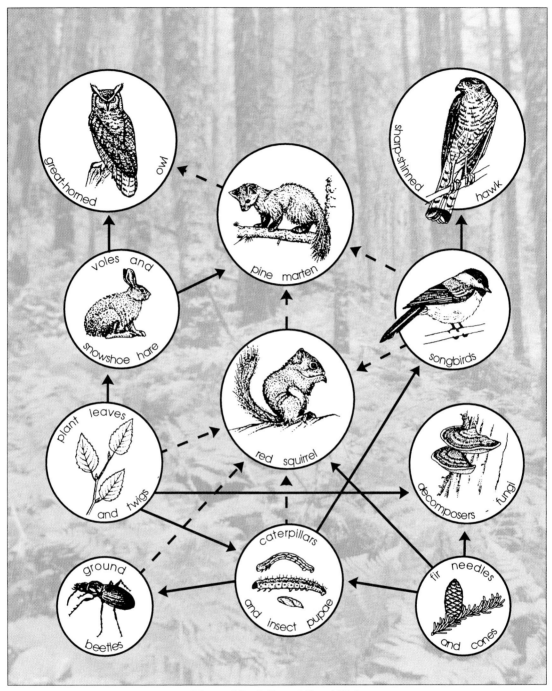

Figure 1.2 - A Forest Food Web.
Solid lines represent main food sources and dashed lines represent food occasionally eaten by the consumer.
(Use this figure to answer the questions on page 32.)

Analysis:

Use Figure 1.2 to answer the following questions:

10. *Identify the producers, primary consumers and secondary consumers.*

11. *What organisms are omnivores?*

12. *What is the advantage of an organism relying on more than one source of food?*

13. *If snowshoe hare were removed from the forest food web, which organisms would be directly affected? Which ones would be indirectly affected?*

Energy Flow and the Earth's Great Cycles

Earlier in this chapter we looked at the earth as one big ecosystem and how homeostasis is necessary to keep it working smoothly. Now we will look more closely at the earth's great cycles—water, carbon/oxygen, nitrogen, minerals and energy flow—all of which are essential to the homeostasis of the earth.

Energy Flow

Close your eyes and picture this... Deep, powdery snow deadens all sound in the frosty woods. A hare with huge, snowshoe feet manages to move around without sinking too far into the snow while it cautiously feeds on young birch twigs. A slight breeze through the dark trees creates just enough noise to cover the whisper of a lynx as it gets within pouncing distance of the hare. Every fibre of the hungry grey cat is intent on making a meal of the smaller but incredibly nimble forest dweller. The lynx moves and the hare reacts — but it's too late. An ancient forest relationship between predator and prey continues, this time favouring the lynx, but more often than not the hare is the winner.

The energy that allows the lynx to catch the hare originally came from the sun. Energy is the ability to do work. You cannot see energy or carry it in a bucket, but you can see the work energy does. Energy from the sun is called solar energy, and it is the source of energy for most life.

Green plants, such as the birch which feeds the hare, convert solar energy into chemical energy through a process called **photosynthesis**. During photosynthesis, chlorophyll uses solar energy to combine carbon dioxide from the air with water molecules from the soil to form

Energy from the sun eventually reaches the lynx allowing it to capture its favorite food—snowshoe hare.

Cellulose: a sugar compound that forms the walls of plant cells.

Respiration: the process by which plants and animals release chemical energy to do work. Work can be growth, reproduction or movement.

Photosynthesis: the process by which plants convert light energy to chemical energy by combining carbon and water to form sugar.

32

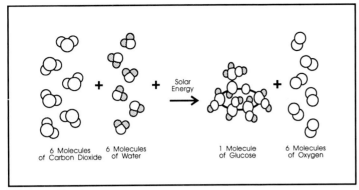

Photosynthesis

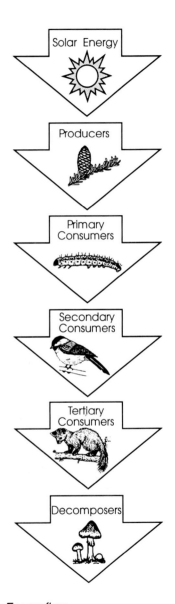

Energy flow

a simple sugar called glucose. Oxygen is also produced through this process and is released into the atmosphere.

Glucose is a source of chemical energy. It may be used directly by the plant or it may be stored. Unused glucose is stored as starch or joined into long chains of **cellulose**, the material that forms the walls of plant cells and creates the basic structure of the plant. The glucose could also be combined with other nutrients like nitrogen, phosphorous and sulphur to form new molecules like proteins or oils.

It is this cellulose and the other material in the plant cells that are the basic food of the hare. The hare's digestive system breaks down the cellulose into glucose molecules, which are broken down further in the cells of the hare. The energy released from breaking down the glucose molecules is used to keep the hare warm, to allow it to move about, to operate its organs, and to build new protein and fat molecules. The process of breaking down glucose to release energy is called **respiration**. Carbon dioxide and water are two of the waste products of this reaction.

When the lynx eats the hare, the energy that temporarily gave life to the hare is used as energy for the lynx. Eventually the lynx will die. When it does, it will provide energy to all the decomposing microbes, insects, and scavenging animals that feed on its carcass.

The energy that first comes to the forest as sunlight moves through all the plants and animals of the forest and is released, bit by bit, back into the environment in a never-ending flow. Energy in all ecosystems follows a very similar path.

33

14. *What is energy?*

15. *Discuss the statement, "Energy flows through the biosphere rather than cycling through it."*

16. *How is energy lost as it travels through an ecosystem?*

17. *How are photosynthesis and respiration related?*

A Closer Look: a species that eats its own feces

It may seem like a wild idea to us, but to the snowshoe hare, it's just business as usual to eat its own droppings.

The snowshoe hare feeds entirely on plant matter. All plant cells are surrounded by cellulose, and in order for the hare to get the benefits of the plant's proteins and fats, it must first digest the cellulose. To do this, it must digest its food twice.

The hare will eat its droppings after the first pass through, and these are stored for a time in its stomach. They eventually mix with fresh food material and are passed on to the small intestine where most of the digested nutrients are absorbed. These nutrients then become a useful source of energy for the snowshoe hare.

The next time you see a pile of fairly dark brown, small round pellets, take a

closer look. They are made up of quite dry waste material, like sawdust in appearance. You have encountered the droppings of a snowshoe hare—second time through!

The Water Cycle—
What Comes Down Must Go Up

Consider for a moment one molecule of water pulsing through your blood stream right now. Some time ago, it might have fallen on the slopes of the Andes as rain. Eventually it could have reached the mighty Amazon River where it passed through water plants, small fish or even the mysterious Amazonian Dolphin before it flowed into the vast Atlantic. From there it could have evaporated into the air to be blown as a storm cloud up the coast of North America, only to fall again onto the land near your home. Then it could have drained into your well or the water supply of your community to end up in your drinking glass. This is one example of the water cycle which links freshwater systems, the oceans and all life (see Figure 1.3).

Energy from the sun causes water to evaporate into the atmosphere. This moisture later returns to the earth

Water Cycle:
1. Evaporation from ocean.
2. Evaporation from soil.
3. Evaporation from fresh water.
4. Water loss from plants and animals.
5. Precipitation into the sea.
6. Precipitation onto the land.
7. Evaporated water carried inland.
8. Surface runoff.
9. Ground water seepage.

Figure 1.3 - The global water cycle. (See notes in margin to explain numbers.)

No matter what state it is in, all the water we have now is all we will ever have.

Imagine!

You are made up of about 65% water.

The large drop represents all the earth's water. The medium-sized drop represents the earth's fresh water. The tiny drop represents the water available to serve the needs of people.

as rain or snow. Thus it is the sun that drives the water cycle.

Without water there would be no life on earth. However, most of the earth's water (97.5%) is in the oceans, which cover about 75% of the earth's surface. Only 2.5% of the earth's water is fresh water and the majority of this is frozen in glaciers or polar ice caps or is deep underground. Only 0.01% of the earth's water is available to serve the needs of people. That doesn't leave much room for improper use.

If water were not so common, we could consider it a wonder substance. One of its most important characteristics is that it will dissolve more substances than any other liquid. In fact, water in a pure state is not found anywhere in nature.

Water dissolves minerals from the earth's rocks and makes them available to plants and animals. It moves minerals and other chemicals through ecosystems. It transports chemicals within the bodies of living organisms to carry nutrients to cells and to rid them of waste products. The chemical processes which keep all creatures alive occur in a watery substance contained in living tissues. In fact, most living creatures are composed of between 65% and 80% water.

Water absorbs amazing amounts of heat, which is distributed around the globe and released slowly. Because the oceans are so vast and can contain so much heat they stabilize the temperature of the earth. This accounts for the relatively small temperature range found on the earth's surface.

Unlike most other substances, water molecules may be found in nature in all three states of matter—solid, liquid, or gas. They are most active as a gas in the atmosphere, usually remaining in that state for about nine days. In contrast, they might be part of a polar ice cap and remain in that state for 10,000 years. (In fact, many of the icebergs that drift past our own coastline contain water that has been locked in an icy state for that long.) No matter what state it is in, all the water we have now is all we will ever have. It moves endlessly through our skies, rivers, oceans, plants and animals. The water cycle is a major part of the earth's homeostasis (see page 21).

Activity: Cycles in a Small World

A terrarium or an aquarium is an example of a closed system—a miniature world in which plants and animals can live and breathe without requiring input from outside their glass containers. Try setting up one of these in your classroom. Through careful analysis you should be able to trace energy flows, and see examples of water and nutrient cycles in action. Can you  identify the consumers and producers? What is the main source of energy driving the system? What changes would you expect if all green plants were taken out? What would happen to the system if the seal were removed?

Analysis:

18. *What drives the water cycle?*

19. *Outline positive and negative ways in which people use the water cycle.*

20. *Water is often referred to as the 'universal solvent'. Explain this in terms of the role it plays in living organisms.*

Nutrient Cycles

Since its beginning, the earth has had all the nutrients it will ever have. They are constantly recycled through our ecosystems. Microscopic organisms in water and soil are vital to these cycles. They release the nutrients that have been captured by plants and animals to make these vital chemicals available to new plants and animals. Without microscopic organisms, all life on the earth would wither and die.

For hundreds of thousands of years, the cycling of most nutrients has been rather slow. Very recently, during the last 100 years or so, the speed with which nutrients have flowed from the land to the sea has increased greatly. We have not yet found a way to reclaim these nutrients from the sea as quickly as they are lost to it.

The most important nutrient cycles are the carbon/oxygen cycle, the nitrogen cycle and the phosphorous cycle (see Figure 1.4, pages 42-43).

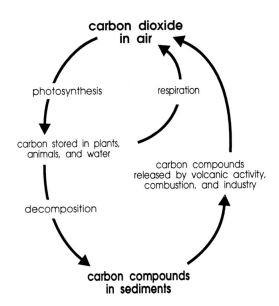

Carbon cycle

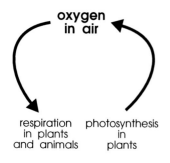

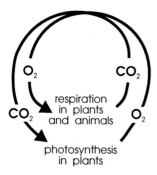

Oxygen cycle

Imagine!

Although the atmosphere is now about 21% oxygen, it was not always at this level. The amount of oxygen increased as simple plant organisms evolved and began to produce oxygen through photosynthesis.

> **Think About It:**
> *What might account for the increase in the loss of nutrients to the sea in the last 100 years?*

Carbon/Oxygen Cycle - The Breathing Earth

The cycling of carbon and oxygen are intimately linked. Both carbon and oxygen are present in the atmosphere and in the hydrosphere.

All molecules that make up living organisms have carbon atoms or chains of carbon as their framework. Carbon becomes part of plants through photosynthesis. Using energy from sunlight, plants turn carbon dioxide from the air, and water from the soil, into glucose, which can be turned into other compounds. When the carbon dioxide is converted by plants to glucose, oxygen is released to the atmosphere. The oxygen released by plants is the source upon which all animals rely.

Whereas plants use the sun's energy to create molecules out of carbon dioxide and to release oxygen, animals do the reverse. They combine oxygen with glucose molecules to release energy. In so doing, they release carbon dioxide back to the atmosphere. This is called respiration and it is going on in your cells right now as you read this book.

Mammals, such as people, absorb oxygen through their lungs, while other creatures, like fish, absorb it directly from water through their gills.

Plants also respire to release their stored energy for growth and reproduction, but they can do this during the day or at night when they are not busy capturing sunlight.

The carbon-containing molecules produced by plants are passed on to animals and decomposers through food chains. Eventually, carbon is released back to the atmosphere or water through the decomposition of living things.

Carbon is also released by natural fires and volcanic activity, as well as the burning of fuels such as wood, coal, gas and oil. Coal, natural gas, and oil all contain carbon since they are the transformed remains of plant life that existed millions of years ago. The plants were changed into either coal or oil, depending on the type of plant material and the conditions of heat and pressure to which they were exposed. Since fossils are the preserved remains of plants or animals, the term *fossil fuels* is used to describe coal, oil and gas.

In the carbon cycle, the oceans serve as an immense storehouse for most of the earth's carbon — holding 50 times as much as the atmosphere. Oceans also serve as a shock absorber for the atmosphere by helping to smooth out or absorb excess carbon dioxide from the air, then slowly releasing it back again. However, we are burning coal, gas and oil at such a high rate that we are overloading the atmosphere with carbon dioxide faster than the oceans can absorb it. This problem is the basis for global warming, which we will look at in more detail in Unit 3.

Nitrogen Cycle

About 78% of the atmosphere is nitrogen. All living things need nitrogen to make proteins. Yet, even though it is so abundant in the atmosphere, neither plants nor animals can use nitrogen directly. Nitrogen usually has to be combined with oxygen as a nitrate before plants can absorb it. Lightning can cause nitrogen and oxygen in the atmosphere to join as nitrogen oxides, but this is not the only way it can happen. Through a process called nitrogen fixation, many bacteria and blue-green algae can combine them as well. Organisms that can do this are called nitrogen-fixing organisms. Some of the nitrogen-fixing bacteria live in the root nodules of plants such as peas and beans (members of the legumes family). These bacteria are called Rhizobium bacteria and are very important in the nitrogen cycle. Some nitrates also come from the erosion of nitrate-rich rocks.

Plants absorb nitrate from the soil and incorporate it into their tissues as part of protein molecules. Animals ingest nitrogen when they eat plants or other animals and use it to make more protein molecules. When plants and animals die, the proteins are converted to ammonia by bacteria.

Before [the industrial revolution] the level of carbon dioxide in the atmosphere never rose above 280 parts per million. By the early 1900s...those levels began to climb. At the end of 1989 they were at 345 parts per million, and by 2070 they are predicted to climb to more than 560 parts per million.

Will Steger

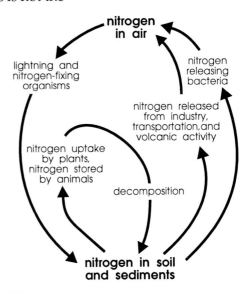

Nitrogen cycle

Plants readily take up ammonia, which is the most common source of nitrogen in the soil. Nitrate is also present but it leaches out easily, making it less available to plants. The conversion of nitrate to ammonia by bacteria is an important process for maintaining soil fertility. Bacteria also release nitrogen from some of the nitrites and nitrates back to the atmosphere.

Humans have influenced the cycling of nitrogen by manufacturing large quantities of nitrates and using them as fertilizers. Much of this nitrogen is lost from the land through soil erosion and ends up in the water systems, causing serious pollution. We also release nitrogen oxides into the air when we burn fossil fuels. This contributes to acid rain, which will be considered in more detail in Unit 3.

Mineral Cycles

Minerals enter ecosystems after they are released from rocks, become part of the soil, and are absorbed by plants. Two of the most important minerals to living organisms are phosphorous and sulphur. Phosphorous is one of the rarest and most important nutrients. All living creatures use it in their metabolism and it is a major part of cell membranes, bone, and teeth. The phosphate cycle is simple and involves only the soil, water and living components of an ecosystem.

Plants take in phosphates directly from the soil or water. Animals get their supply by eating plants or other animals. Excess phosphate is excreted in the animals' urine and returned to the soil or water to complete the cycle. Phosphorous enters the atmosphere only as dust, an industrial byproduct.

The amount of phosphorous in aquatic habitats limits plant growth. We add significant amounts of phosphate to aquatic systems because it is a major component of fertilizer and detergents which enter our water systems as runoff and sewage. This can stimulate plant growth un-

---

Imagine!

Farm animals in Canada produce 322 million litres of manure per day. This material contains immense quantities of nitrogen, phosphorous and potassium, which, if it were in the form of equivalent chemical fertilizer, would be valued at about $900 million per year.

Environment Canada

Mineral cycle

naturally. When this plant growth dies, decomposing organisms feast on the abundant food supply. As this occurs, the decomposers use large amounts of oxygen from the surrounding water. This eventually leads to oxygen 'starvation' in the river, lake or pond.

Sulphur is an important part of proteins. The sulphur cycle is similar to the phosphorous cycle in that the main source is from rocks. It becomes part of the soil through erosion and is then available to plants and animals. Unlike phosphorous, high levels of sulphur are present in fossil fuels, especially coal. When these fuels are burned, sulphur dioxide is released to the atmosphere and is a major cause of acid rain.

Although we have looked only at phosphorous and sulphur here, other mineral cycles are also important. These include the cycles of calcium, sodium, potassium, and magnesium.

From space we can see and study the Earth as an organism whose health depends on all of its parts.
World Commission on
Environment and Development

Analysis:

21. List three nutrients that are essential to life. What role do they play?

22. What might account for the loss of nutrients from a forest ecosystem?

23. As a byproduct of respiration, carbon dioxide is released to the atmosphere. Outline two ways that carbon dioxide is removed from the atmosphere.

24. Farmers often plant crops of clover or beans every few years. Why?

25. Many detergent companies now manufacture phosphate-free detergent. Why?

26. Research the calcium cycle, or another mineral cycle, and develop a flow chart showing its path through an ecosystem.

Pyramids of Life

To get a clearer picture of the connections between living organisms in an ecosystem, think of these connections as feeding levels. The green plants or producers form the bottom level; primary consumers that feed on the green plants form the next level; secondary consumers, or primary carnivores, that feed on primary consumers are next; and tertiary consumers that feed on other carnivores are at the top. A more complex ecosystem may have more than four levels, but rarely will there be more than five. We call each of these levels a **trophic level**.

Trophic level: position in a food or energy chain determined by the number of supporting levels.

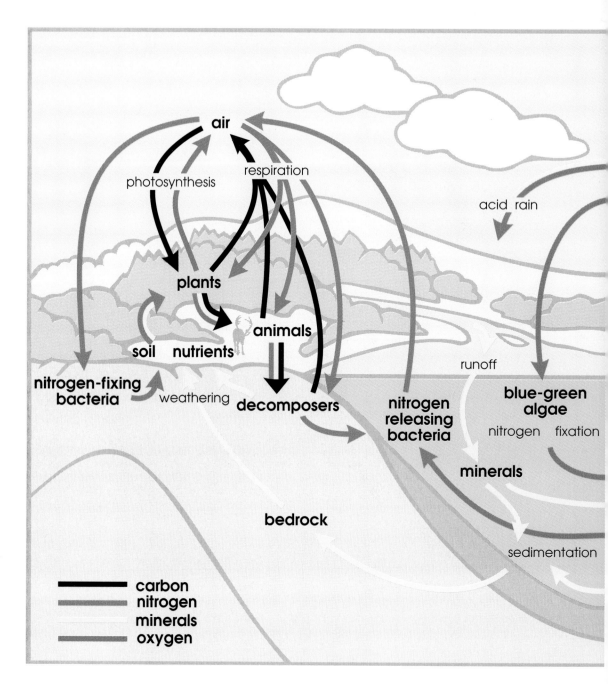

Figure 1.4 - The earth's major

The earth's homeostasis depends on the continuous recycling of carbon, oxygen, nitrogen and minerals now are all the resources we will ever have.

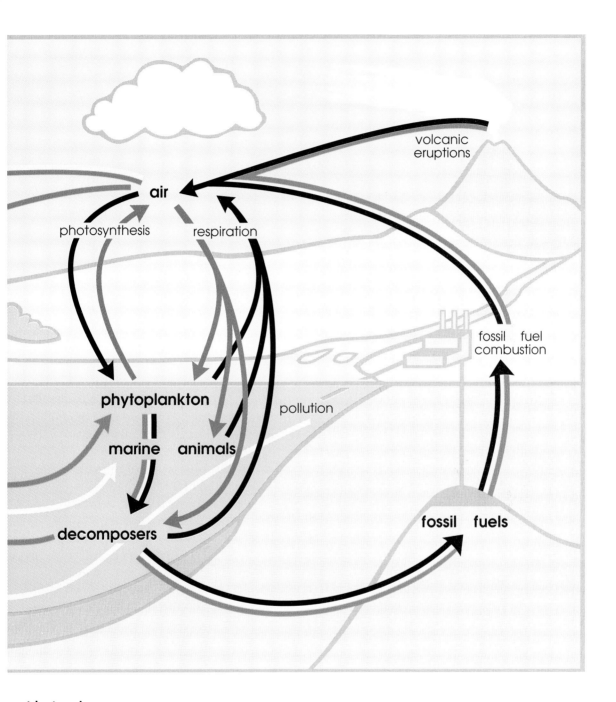

nutrient cycles.

through the animals, plants, water, soil and atmosphere. Except for solar energy, all the resources we have

Biomass: mass of one group of living organisms.

Joule: the amount of energy expended when a force of one newton is exerted through a distance of one metre.

Pyramid of Numbers

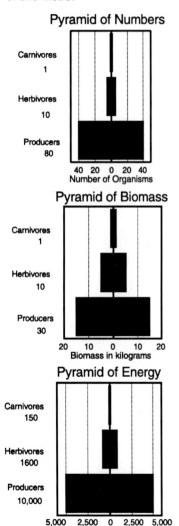

Carnivores
1

Herbivores
10

Producers
80

40 20 0 20 40
Number of Organisms

Pyramid of Biomass

Carnivores
1

Herbivores
10

Producers
30

20   10   0   10   20
Biomass in kilograms

Pyramid of Energy

Carnivores
150

Herbivores
1600

Producers
10,000

5,000   2,500   0   2,500   5,000
Energy in kilojoules

If you draw each of the trophic levels to represent the number of organisms in that level, all the levels together will usually form a pyramid. Ecologists use three different pyramids to understand ecosystems. The first deals with the actual number of organisms at each level; the second deals with the **biomass**, or weight, of all the organisms at each level; and the last deals with the amount of energy at each level.

In any ecosystem, it usually takes many plants to feed a lesser number of primary consumers, which in turn feed even fewer secondary consumers. For example, it takes many hares to keep one lynx alive for a year. Therefore there are always many more hares in a forest than lynx. As you move up through the trophic levels of an ecosystem from plants to carnivores, the number of organisms usually decreases. This reduction in the number of organisms at each level is what creates the pyramid shape.

Not all pyramids of numbers follow this pattern, however, especially when the plants are large trees. For example, a few fir trees would feed many thousand spruce budworms (primary consumers), which in turn might feed twenty insect-eating birds (secondary consumers). In this case the pyramid has a very small base (the few fir trees).

However, if you draw the trophic levels of the same section of forest as biomass or weight, rather than as the number of organisms, the pyramid will have a more characteristic shape, with a broad base and a narrowing top. This is because the trees contain a great deal of matter (biomass) though they are few in number.

The third type of pyramid used by ecologists deals with energy. The units used to describe a quantity of energy are **joule**. You can convert a biomass pyramid to an energy pyramid by assuming there are about 16 joules of energy in each gram of biomass. Remember from the discussion of energy flow that plants and animals use large amounts of energy for their own life processes such as growth, reproduction, movement, and maintenance of a stable temperature.

Energy that is used at one trophic level is not available to the next. Since energy is lost between each level of the ecosystem, the amount stored in the green plants must be much greater than the amount stored in the bodies of the primary consumers. Only about 10 % of the energy contained in the cells of producers ends up in the cells of the primary consumer, and only about 10% of that energy reaches the cells of the secondary consumer. Because so much energy is lost (about 90%) between each trophic level, it is rare for an ecosystem to have more than five levels. There is simply not enough energy to support more than that.

Think About It:

It is important to know that all the pyramids for any ecosystem can change shape with time. If too many herbivores destroy the food supply, what would be the new shape of the biomass pyramid? Could this last?

Imagine!

An animal like a fox can be on different trophic levels depending on what it eats. When it eats berries it is a primary consumer, but when it eats a shrew, it is a tertiary consumer.

Ecologists use the shape of pyramids and the number of trophic levels to compare one ecosystem to another and to predict changes within ecosystems. Any ecosystem with very large values for biomass and energy would be a highly productive one.

Analysis:

27. *What is a pyramid of numbers?*

28. *What is one of the difficulties in using a pyramid of numbers to describe an ecosystem?*

29. *Distinguish between a pyramid of numbers and a pyramid of biomass.*

30. *What is the major factor that determines the number of trophic levels in an ecosystem?*

31. *Why is it difficult to understand and predict all the connections in an ecosystem?*

32. *Can you give two examples where human actions or developments were seriously harmful to an ecosystem?*

Last Thoughts

The ecosystems of the earth, the water and nutrient cycles, and the flow of energy create a oneness within the biosphere. All creatures are dependent on and contribute to the smooth working of the whole. As one of the powerful creatures of the earth, people are in a special position. We can cause great damage to the earth or we can help it function smoothly.

This gives each of us personal choices. With each choice comes a responsibility to decide if the earth will be better or worse as a result of that choice. Right now the choices you make may feel insignificant, but they will help shape you into a future citizen of the earth. For the rest of your life you will continue to make everyday choices about the things you buy, the way you work and play, the way you raise your family, what you support and what you oppose. Many of you will become influential decision-makers whose choices will affect the lives of hundreds or even thousands of other people, as well as the other creatures sharing this earth. Our record over the past 50 years or so has been poor. Changes are essential for the earth to maintain its homeostasis. What choices are you going to make?

CHAPTER 2
The Ecosystems of Newfoundland and Labrador

I walked for an hour down, down from the hill, down through swampy valleys, down through scattered patches of spruce, up into ridged hills... Despite the rain and the mist, I saw the configuration of the coastline...then I began to understand the scope and depth of the land.

Franklin Russell
The Secret Islands

What You Will Learn:

- the main ecosystems in Newfoundland and Labrador—wetlands, freshwater systems, oceans, boreal forest, and barrens;

- a closer look at one specific kind of wetland ecosystem—bogs—how they are formed, what they look and feel like, their function in nature, the way people use bogs, and the effect we have on them;

- communities of living things as changing systems over time;

- population density and control—ways of measuring populations of living things and of using this information to assist in maintaining a healthy balance within the natural environment;

- ecoregions—defined areas of ecosystems.

Community: an association of organisms living in a common environment.

Wetlands
The Sponges of the Earth

Wetlands are just what the name suggests—wet lands; areas of land saturated with water. Wetlands support life that is adapted to a wet environment. Picture giant sponges on the earth's surface, soaking up and storing water from the land and air. But wetlands are more than sponges; they are among the most important and productive groups of natural communities on earth!

There are five classes of wetlands: *marshes, swamps, shallow water ponds, bogs* and *fens*. Bogs and fens are types of peatlands. Each one is a unique system; a self-supporting **community** of plants and animals interacting with one another and with the nonliving environment—an *ecosystem*.

Peatlands develop in poorly drained areas, while marshes and swamps occur in areas that are periodically flooded, such as floodplains, riverbanks and estuaries. Ninety-nine percent of the wetlands in Newfoundland and Labrador are peatlands. Surveys have shown that 11% of the land surface of the island of Newfoundland is covered by peatlands. Most of these peatlands occur on coastal lowlands and the south-central inland plateaus. In Labrador, peatlands cover an estimated 19% of the land surface and are concentrated mainly in the western and southern regions.

Types of Peatlands

Peatlands are so named because of their thick layer of **peat**, a layer of **organic soil** that is at least 40 centimetres thick. There are two distinct types of peatlands—bogs and fens. However, few peatlands are strictly of one type or the other. The most important difference between fens and bogs is that bogs are very poor in nutrients, which are brought in mainly by precipitation (rainfall), while fens are richer in nutrients, brought by water seeping in from **mineral soils** nearby. Bogs have much higher acidity levels than fens.

Because there are different kinds of bogs, their appearance can vary. Bogs are relatively flat, open areas. The surface is a continuous carpet of sphagnum moss interrupted by hummocks and hollows. The slightly raised hummocks are covered with shrubs and perhaps a stunted tree or two, while the lower, wet hollows have more **sedges** and grasses. Scattered here and there are small pools with black peat showing around the edges.

A fen has lots of shrubs but looks more meadow-like than a bog because there are more sedges and grasses and they grow more tightly together. Trees can grow better in fens than on bogs. In fact, some fens might be completely forested. Scattered pools often break the surface of a fen, but in others the water, although only centimetres from the surface, may not be visible. Both fens and bogs can vary greatly in size.

Bog

Fen

Peat: the remains of partially decayed plants.

Organic soil: soil which contains carbon because it was once part of living tissue. Organic matter is a source of food for bacteria and is usually combustible.

Mineral soil: soil composed primarily of minerals, in contrast to organic soil which is composed of plant and animal remains.

Sedges: plants similar to grasses but with three-sided stems instead of flat stems like grasses.

Climax community: the final stage of successional change. A stable community.

Succession
A Chain of Changes

As environmental conditions change, one plant community can be replaced by another. This gradual change from one plant community to another is called *succession*. Since each area is influenced by a different set of environmental factors, the path and outcome of succession may be different for each location. Succession proceeds until a more stable **climax community** is established.

Infilling: the change from open water to a bog caused by plant life growing in from the edges.

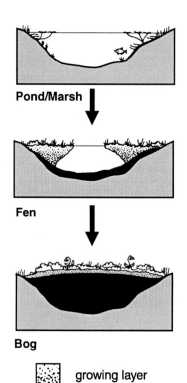

Pond/Marsh

Fen

Bog

growing layer

peat

mineral soil

Figure 2.1 - Bog succession.

One type of succession is pond infilling, which leads to the formation of a bog. As lake sediments and organic matter build up in a pond, aquatic and marsh plants begin to grow along the edges and slowly spread towards the centre. As plant growth continues, the flow of water becomes restricted. Slow-moving or standing water contains less dissolved oxygen. Organisms that normally break down dead plant matter cannot survive in this low oxygen environment, thus less decomposition occurs and fewer nutrients are released. As peat accumulates from decaying sedges and grasses, the wetland type changes from a marsh to a fen. As long as there is a balance between plant production and plant decomposition, the fen community will remain fairly stable. However, if the rate of plant production exceeds the rate of decomposition, more peat will build up and the surface of the fen will rise. The surface may rise to a point where the nutrient-rich and oxygen-rich seepage waters no longer reach it. Now precipitation is the only source of nutrients and the fen plants will be replaced gradually by bog plants (see figure 2.1).

In some regions of Canada, bog succession will continue to a climax fir or spruce forest. But the cool, wet climate of Newfoundland and Labrador prevents the bog surface from drying out enough to allow the forest species to invade. Here, a bog is the final stage of pond **infilling**, the climax community.

All ecosystems experience some form of succession.

Life in Peatlands
Making Connections

Plantlife

In peatlands, the dominant plants are shrubs and mosses, especially *sphagnum moss*. In nutrient-poor bogs, this amazing sponge-like moss provides the foundation or platform upon which all other plants grow. It is especially adapted to hold large quantities of water. In fact, native peoples have used sphagnum moss in diapers! Where the surface of sphagnum rises in hummocks above the water table, the roots of other plants can take hold. **Heaths**, such as bog laurel, Labrador tea and bog rosemary manage to thrive there, as do caribou lichens and bakeapples. A few stunted trees, such as scrubby larch and black spruce find the growing conditions barely adequate.

In contrast to plants living on mineral soil, those in a bog must survive without a rich supply of nutrients. Some plants, such as sundews and pitcher plants, have adapted to this environment in special ways. Since fens

Heaths: plants that are members of the blueberry family. Most have evergreen leaves.

A Closer Look: carnivorous plants

Carnivorous plants are adapted to live in nutrient-poor areas such as peatlands. Although these unique plants manufacture most of their food through photosynthesis, they supplement their diet by capturing and digesting tiny insects. Two common carnivorous plants are the *pitcher plant* and *sundew.*

Pitcher plants have hollow pitcher-like leaves filled with rainwater and digestive juices. Insects, attracted by nectar to the showy leaves, fall from the smooth lip into the pitcher. Downward pointing hairs inside the leaf prevent insects from crawling out, causing them to drown. The drowned insect is digested by enzymes and bacteria.

Sundews trap insects with the use of special gland-tipped hairs. These hairs secrete a sweet-smelling liquid to attract insects and a sticky substance to hold them. Once an insect is stuck, the leaf's outer hairs bend inward, pinning down the prey. The insect is then digested by enzymes secreted by the glandular hairs.

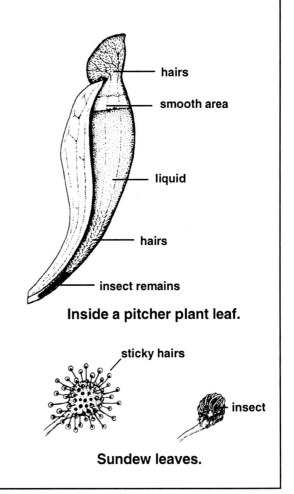

Inside a pitcher plant leaf.

Sundew leaves.

Dragonfly

Sphagnum

have more nutrients than bogs, you will find there more sedges and grasses as well as plants like buckbean, bottlebrush and dwarf birch. Both bogs and fens also have their own specialized and beautiful orchids.

Wildlife

Many kinds of wildlife make their homes in bogs and fens. Anyone who has been in a bog during a warm day in June knows that the most common organisms inhabiting bogs are insects and spiders. Mosquito larvae hatch in the shallow pools. Dragonflies and damselflies dart through the air feeding on other flying insects. Spiders, silent by their webs, await their prey. Small flies, bees and butterflies flit from flower to flower sipping sweet nectar and pollinating the plants as they go. Some of these insects are found only in bogs or fens. The pools of bogs and fens provide a home for the few amphibians found in our province, such as the green frog.

As in every other type of natural community, the animal life in our bogs and fens can be classified as either *herbivores, omnivores, carnivores* or *decomposers*. All are linked together in complex *interdependencies*. Figure 2.2 shows some of the relationships that are characteristic of most bogs.

The meadow vole, found on the island of Newfoundland and in Labrador, is a small mouse-like mammal that tunnels through the grasses and shrubs feeding on a variety of plants. In Labrador, two other small mammals with similar habits use bogs and fens—the red-backed vole and the bog lemming. Together with the snowshoe hare, these animals provide the main food for some of the predators, or hunters of our bogs and fens. Among these predators are the weasel or ermine, the fox, the lynx, the short-eared owl and the northern harrier (also known as the marsh hawk). Other birds that use bogs and fens are the Canada goose, the swamp sparrow, willow ptarmigan (or partridge) and the snipe.

The largest animals that frequent bogs and fens are moose, caribou and bears. These animals are attracted to peatlands because they provide easy travel routes, a number of plants they can eat, and breezy areas where they can get some relief from biting flies.

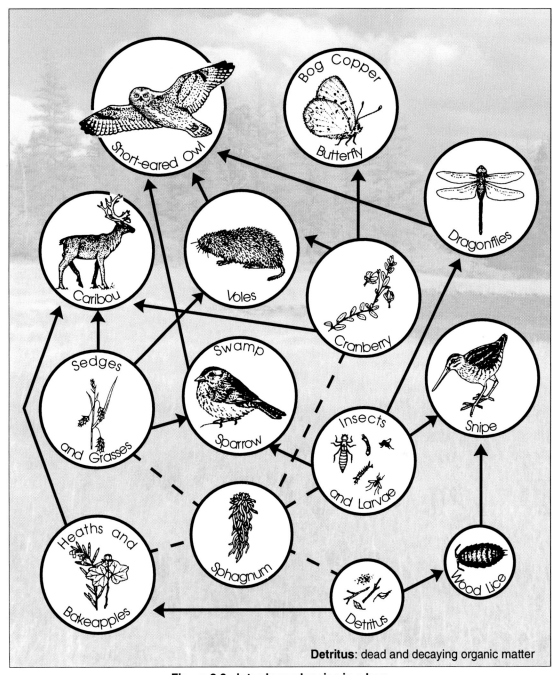

Detritus: dead and decaying organic matter

Figure 2.2 - Interdependencies in a bog.

A food web typical of a bog ecosystem shows the order in which smaller organisms are consumed by larger ones. For example, sedges and grasses are eaten by meadow voles, which in turn are eaten by the short-eared owl. All of these organisms rely on the sphagnum moss which forms the platform for all other life in a bog. (Solid lines represent energy flow; broken lines represent other dependencies.)

Bogs are Important, Naturally

American Bittern

Habitat: a place that provides all the food, water, shelter and space necessary for a particular organism to thrive.

Although peatlands play an important role in nature, they are not always valued. For years people have treated peatlands as expendable wasteland, a problem to overcome when constructing a new road or building. In many homes throughout the province, stories are told of the loss of the horse, cow or sheep after it became mired in a bog hole. People have drained peatlands, torn them up with all-terrain vehicles, flooded them and dumped their garbage in them. Even our use of language shows that we think of bogs as barriers to progress; hence the saying, "bogged down."

So, what good is a bog? Just like sponges, bogs and fens have a tremendous ability to store water and release it slowly. This characteristic helps prevent flooding and erosion. Some peatlands improve water quality by straining out suspended particles in their dense and matted vegetation. Peatlands also provide water, food, nesting and resting areas for a variety of wildlife. In other words, they are important **habitats**.

Each of these functions should remind us of the importance of protecting all wetlands, not simply because they are special in themselves, but because they are connected with so many other parts of our environment.

A Closer Look: bog people

In recent years, scientists found the bodies of several people who had been buried in a bog in Denmark for two thousand years. The remains were remarkably well preserved, with hair, skin and even fingerprints still intact. The fact that these remains were preserved so perfectly tells us much about the characteristics of the bog. The waterlogged peat and lack of oxygen prevented bacteria and other decomposers from rotting the flesh. This process was helped by the high *acidity* of the bog which also softened the bones and turned the skin dark brown.

Using Our Peatlands

Are bogs more of a nuisance than a blessing? While we could debate any reply to this question, we can agree that they are a source of great beauty and inspiration. Nothing can quite match the sight of orange deergrass early on a still, fall morning as the sun begins to burn off the ground-clinging mists. And we know too that our peatlands provide homes to a large number of plants and animals, some of which end up on our dinner tables.

We have known for some time that our peatlands are a resource. With proper drainage and fertilization, peatlands can provide land for agriculture and grazing as well as forest lands. Today many families use peatlands as rich and productive plots for growing their vegetables. Mined peat is used as a growing medium and soil conditioner in gardens and potting soil.

Peat can be burned to produce heat. In some parts of the world, large-scale peat harvests have been used to fuel hydroelectric plants.

If peatlands are a resource, are they **renewable** or **nonrenewable**? The answer depends on how they are used. Peat forms over thousands of years. If people harvest it in large quantities and destroy the living surface of a bog, it cannot renew itself. When all-terrain vehicles are driven over peatland, they can crush the mat of living plants on top and cause the root systems to separate. This results in deep scarring of the bog. If our peatlands are to be developed, they must be carefully managed.

Other peat-rich countries such as Ireland, Finland, Russia and the United States have been making extensive use of their peat resources. Because our province is relatively late starting, we can benefit from the experiences of these countries. We can learn from the mistakes they may have made. If we choose to we could develop some of our peatland resources without forgetting to consider the benefits they provide, both for us and future generations. In other words, we can apply the principles of sustainable development.

Imagine!
Coal is formed from peat that is thousands of years old and highly compressed by changes in the earth's crust.

Renewable resource: living resources that can renew themselves when conditions permit.

Nonrenewable resource: nonliving resources such as rocks and minerals which cannot renew themselves.

Peat harvesting in Trepassey

Aerial view of ATV damage

55

The Economic Value of an Ecosystem

Bakeapple

Bog wildlife

How can we determine the economic value of an ecosystem? This is one of the most difficult questions in environmental science. Consider the value of a wetland. Should it be based on how much you could sell the products of a wetland for? If you used this as a means of determining value, how would you value all the parts of an ecosystem that no one wanted to buy? Should the value be based on how much it would cost if you had to replace a wetland after it was destroyed? Should it be based on the amount of time people spend using wetlands?

None of these methods is entirely satisfactory but they are all used to estimate the economic value of ecosystems. Table 2.1 shows a value for Canadian wetlands based on combining the value of people's time using wetlands with their replacement value.

Type of Use	Estimated annual value in millions of $
Photography, bird watching, tourism, education	3,000.0
Protection from floods	2,700.0
Water purification	1,350.0
Recreational and commercial fishing	1,278.4
Peatland forestry	525.0
Hunting and trapping	304.3
Aquaculture (fish farming) and agriculture	102.8
Peat for energy and horticulture	47.8
Wild rice harvesting	7.0
Total Value of Canadian Wetlands	$9,315.3

Table 2.1 - Wetland economic values for Canada.

Activity: Down on the Bog

The best way to learn about bogs, or for that matter, any other ecosystem, is to spend time there! So grab your boots and jacket, take along your environmental journal and field equipment and head for the bog.

Your teacher will provide you with a detailed set of field instructions.

Upon arrival, take some time to sense the unique beauty of this place. You may see, feel, smell and hear things you never noticed before... the squelching, undulating sphagnum moss under your feet; the sounds of birds and other wildlife; the hardy but delicate beauty of the plant life; the colour and the earthy, moist smells of this wet world. This is a rich environment for edible plants, and depending on the time of year, you may find an abundance of bakeapples or cranberries.

1. Jot a few words about this special place in your environmental journal. You might feel like writing a poem or doing a sketch of something that impresses you.

2. List the signs of human impact on the bog. List any signs of other animal activity.

3. Sketch a rough map of your study area. On your map, indicate true north, any streams' or creeks' direction of flow of water and distinctive areas of vegetation.

4. Use the information provided by your teacher to identify at least four plants from your study area.

5. Determine the relative level of nitrogen in a sample of peat.

6. Determine the pH of a sample of bog water.

7. Determine the carbon dioxide concentration in a sample of bog water.

8. Use a sample plot or quadrat to determine the relative population of one species of bog plant.

9. Construct a vertical profile of the peat depth of your study area by running a transect and pushing a long stick into the peat at one-metre intervals.

Analysis:

1. List the five different classes of wetlands, and list two distinguishing characteristics of each.

2. Describe succession from a pond to bog. In what ways do you think people can influence the rate at which succession occurs or does not occur?

3. Give two reasons why animals find bogs so attractive.

4. Distinguish between a renewable and nonrenewable resource. Give one example of each.

5. All ecosystems are of some economic value. Do you think it is necessary to place a dollar value on ecosystems to justify protecting them? What other value might the bog ecosystem hold besides an economic value?

Freshwater

Imagine!

It is estimated that the rivers of the world transport one billion tons of solid debris to the oceans and about 400 million tons of dissolved matter.

Watershed: the entire drainage area of a river and its tributaries.

Most of us have spent a lazy summer evening beside a pond watching the slow rippling circles form on the surface as trout rise to snatch insects. Your fishing line gently arcs out over the water as you hope a trout will take your bait. There are precious few Newfoundlanders or Labradorians who have never felt the peace and satisfaction gained from this use of our freshwater systems.

The freshwater systems we now use began to form about eight thousand years ago, when the last glaciers melted into the North Atlantic Ocean and left behind basins and deep gouges in the land. After filling with water, these basins became our ponds, lakes, rivers and streams. They provide us with a plentiful supply of fresh water, covering about 8% of our land surface—more than in most regions of Canada.

The Churchill River in Labrador represents the largest **watershed** in the province. It takes in 90,000 square kilometres of land and has an average flow of 1,900 cubic metres per second. This volume would be roughly equal to 11 school buses plunging into the ocean every second! Compare this to the largest river on the island, the Exploits, which has an average flow of 300 cubic metres or about two school buses per second.

58

Freshwater ecosystems are called 'open' ecosystems because they are open to changing conditions from the outside. For example, nutrients can enter the water from an overhanging tree or from soil erosion. Materials that enter a freshwater ecosystem usually move with the water, so that what happens upstream will affect events and life many miles downstream.

Ecologists who study freshwater ecosystems are called *limnologists*. They classify fresh water in two major groups—standing waters and flowing waters. Standing waters are ponds, lakes and wetlands, while flowing waters are brooks, streams and rivers.

Of the standing waters, only two types are entirely aquatic: ponds and lakes. Other types, such as the wetlands you have already studied, are partly terrestrial. The distinction between a pond and a lake is not very clear. However, ponds are usually shallow enough that light can reach the bottom in all places. Lakes may have areas that are too deep for light to penetrate to the bottom. Therefore, ponds are generally smaller and shallower than lakes. Because ponds are shallower, they will have vegetation growing throughout them. Lakes may have no vegetation in deeper sections.

Flowing water feeds most ponds and lakes. It starts as small brooks that join to form rivers. As water flows from the brooks to the rivers, the characteristics of the system change. Generally, small brooks begin in higher, cooler country than rivers. Therefore, they are usually cooler, contain more oxygen and are faster flowing than the rivers into which they empty. The bottoms of most brooks are rocky, since most of the finer materials get washed downstream by the rapid flow. Slower rivers will have sandy or even muddy bottoms. The types of plants and animals that live in brooks and rivers will differ as a result. For example, trout prefer cooler water that has more oxygen.

The vegetation of the surrounding land also affects freshwater systems. Root networks of trees and plants will slow the speed with which water seeps through the soil and will help knit the soil together to slow erosion. Many of the nutrients entering fresh water also originate from the surrounding vegetation.

Imagine!

The processes used to manufacture a car require the use of about 500,000 litres of water.

Imagine!

If the total water consumption of Canadians were divided equally among all of us, each person would use about 9,000 litres of water per day. At the turn of the century, we used about 2,400 litres per day.

Imagine!

People in North America are using fresh water from the land twice as fast as the water cycle can replace it.

Brook Trout

Northern Pike

Arctic Char

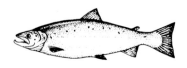

Atlantic Salmon

Geology plays a large part in the refilling of freshwater systems. A Cape Breton fisherman could have been talking about Newfoundland and Labrador when he said, "There's a lot of rock in this land, and a lot of rainwater too. It comes down hard, and it runs off hard." Our shallow soil and open rock surfaces cause rapid runoff and limited water storage. Large limestone deposits on the island's west coast are an exception to this since they have a high capacity for storing water. Nearly 90% of the province's public water supply is drawn from surface water rather than groundwater.

During the glaciation process, most plants and animals that lived here were eliminated. The fish species we now have on the island are those which reached here through the sea. We therefore have two main groups of freshwater fish, both of which can live in either fresh water or salt water. The first group is the *anadromous* fish, which spend most of their lives in salt water but move to fresh water to spawn. These fish include the salmon, brown or sea trout, arctic char, and sticklebacks. The other group is *catadromous* fish, such as eels, which spawn in salt water but spend most of their lives in fresh water.

The freshwater fish of Labrador are more varied than those of the island since they could reach the lakes and rivers from other freshwater systems. Labrador has most of the same fish as the island but it also has pike, whitefish, lake trout, and suckers — species that have not been able to reach the island because of the salt water barrier dividing the province and mainland.

Think About It:

Freshwater systems are often used for the disposal of waste products.

1. *Why do you think we use freshwater systems for this purpose?*
2. *If we continue to use them in this way, what changes can you predict locally, near the disposal site, and further away?*

Analysis:

6. *Why are freshwater ecosystems called 'open' systems? What special considerations have to be made when using lakes, rivers and streams, given that they are open ecosystems?*

7. *What might be two effects of uncontrolled timber harvesting on a watershed?*

The ocean's bottom is a lot more interesting than the moon's behind.

Author unknown

Oceans

Our lives in this province are influenced more by the sea than any other environmental factor. The sea determines our climate, where we live, what we do, and the nature of the ecosystems we are part of. The cold, rich Labrador current carries pack ice, the breeding platform for our seal populations, and icebergs, the dread of fishermen and oil rig workers alike. Its cold waters extend our winters and cause us to think that summer may never come. It determines where and how we build wharves and when ships can come and go around our coast. It is the home of northern cod, which has been the basis of our provincial economy for 400 years. Where the Labrador Current meets the Grand Banks, just off the east coast of Newfoundland, we have one of the world's richest ocean environments.

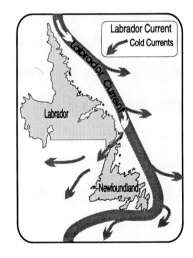

The ocean ecosystem can be thought of as one big ecosystem or a number of smaller ones that are connected to make up the whole. Since oceans are very different from place to place and from top to bottom, scientists who study ocean ecosystems divide them into zones, each of which can be thought of as an ecosystem in itself (see Figure 2.3).

I must go down to the sea again,
for the call of the running tide
is a wild call and a clear call
that may not be denied.

from "Sea Fever"
John Masefield

Plankton: organisms floating in water habitats, including algae, bacteria, fish larvae, small crustaceans and other tiny organisms. Phytoplankton are plants, and zooplankton are animals. They provide food for other ocean animals.

Marine ecosystems are different from those of the land in a number of ways. The seas are salty and very deep, sometimes as deep as 10 km. This gives them about 300 times more living space than is available on land. The land does not move a lot. However, the currents, waves, and tides of the sea move vast bodies of water from one place to another.

Most of the seas' basic nutrients are washed off the land or fall from the atmosphere. That is why the seas are salty. These nutrients are captured briefly by life near the surface of the sea, but most eventually sink to the bottom, where they are quickly diluted by the vast amount of moving water.

The sea may be deep, but sunlight can only penetrate the first 30 to 120 metres. Therefore, plant life that requires sunlight is concentrated close to the surface. Most of these plants are very small, even microscopic, so most of the animals that feed on them are also very small. These tiny plants and animals that are the basis for all other life in the sea are called **plankton**.

As the ocean's plants and animals die, they are either intercepted by scavengers or they sink deep into the ocean. This slow rain of material from the ocean's

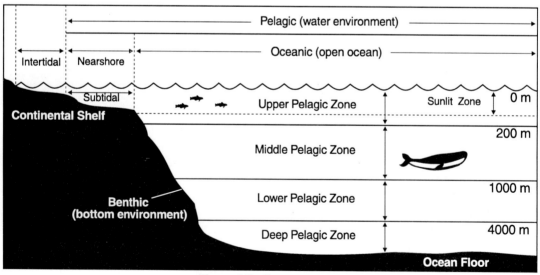

Figure 2.3 - Ocean zones.

surface feeds the mysterious and sparse life in its black depths.

Compared to the land, the sea is not very productive. This is because most life is concentrated near its surface and because the nutrients that support life be-

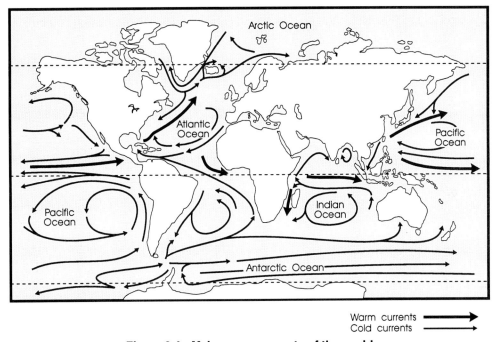

Figure 2.4 - Main ocean currents of the world.

come thinly distributed in the vast volume of water. Constant gravity slowly pulls both living and nonliving substances to the depths where they are out of reach to most organisms. Although the seas are very large, they only produce about one-quarter of the earth's life.

People benefit from the oceans in many ways. The oceans' role in global processes such as the water and carbon cycles is part of what makes human life on earth possible. A significant portion of the world's oxygen is produced by ocean plants, especially phytoplankton. (refer to page 35 to review the role of the oceans). The earth's climate is largely influenced by ocean currents. About 2% of the world's food comes from oceans, providing about 5% of the world's protein needs. Some countries rely even more heavily on fish products, using it to supply

Marine Food Chains
Because the plants and animals at the base of the oceans' ecological pyramid are so numerous and small, there are usually more trophic levels between the producers at the bottom and the carnivores at the top of the food chain. As a result, marine food chains tend to be longer than those of the land.

most of their protein needs. Our shipping industry moves more products from place to place on the surface of the ocean than on any other of the earth's surfaces. These are only a few of the benefits we derive from the oceans, yet people's existence everywhere relies heavily on keeping our oceans healthy.

A Closer Look: life in a tough community

Escaping from a life-crushing force, hiding from enemies, capturing prey, making babies, finding a meal, living and dying. No, this is not a soap opera plot or an Arnold Schwarzenegger movie. This is a day in the life of a puddle of water along any of our rocky coasts.

Tide pools are small pools of water left after the tide goes out, and they are filled with many forms of life. The plants and animals that survive in these pools have to withstand not only the rising and falling tides, but also the harsh pounding waves and pressure from the open Atlantic Ocean.

Cracks in rocks provide a solid place for seaweeds to anchor. These seaweeds, which in many places completely cover the rocks, help to soften the impact of the waves. Animals have developed special shapes and behaviours to protect them from being washed away or crushed. Blue mussels spin a set of elastic threads which they cling to for support. Periwinkles have hard protective shells and if they are pulled off the rock by a

wave, they simply close their 'operculum' (like a trap door) and roll with the waves until things calm down. Barnacles secrete a cement-like substance once they make their home on a rock. Anyone who has tried to scrape them off the bottom of a boat knows how well they stick!

These are only a handful of the creatures you might find in a tide pool. Next time you're walking along the shore, take a closer look.

Analysis:

8. *"Our lives in this province are influenced more by the sea than any other environmental factor."* *Make a journal entry on how your life is affected by the ocean.*

9. *Why should Newfoundlanders be concerned about ocean pollution that is occurring beyond the 200-mile limit and in other oceans thousands of miles away?*

Forests

Though all the forests of this province are part of the boreal forest biome, scientists recognize six forest types and even more subtypes. Each type has different groups of species and each has a different sensitivity to disturbance.

The most common forest type in the undisturbed boreal forest biome is balsam fir. Balsam fir is well known to most of us at Christmas time. This very productive forest is most common in the western region of Newfoundland and the central and southeastern regions of Labrador. It is called a balsam fir forest type because fir is the dominant tree species present. But it is not the only species present. White birch, white spruce and black spruce also live in the balsam fir forest.

If the forest is logged, the balsam fir species will regenerate, but if there is a fire, it will be replaced by black spruce or white birch forests.

Another abundant forest type is black spruce. Wherever you see large stands of black spruce, you are probably looking at an area that was once burned. In Newfoundland, these forests are most common in the dry northern and central areas. They are common

Black spruce - lichen woodland

Caribou lichen

throughout Labrador, except in the far north where they are replaced by tundra. While black spruce is the dominant tree species, balsam fir, white spruce, white birch and trembling aspen are mixed throughout.

One of the most common forest types in Labrador is black spruce - lichen woodland. The lichen in this forest may be known to you as 'caribou' or 'reindeer' moss but it is not a moss at all. Rather, it is a pale, stiff plant that grows in clumps that look similar to pot scrubbers. These lichens are a major food source for caribou. These beautiful forests with widespread trees are practically absent from the island of Newfoundland.

Some other forest types include hardwoods such as birch and aspen, dwarf shrub - black spruce, and black spruce bog. Anyone who has travelled in the country very much has been slowed by the dwarf evergreen growth known as tuckamore, or simply 'tuck'. Tuck is actually spruce and fir whose growth is stunted by high winds or excess moisture in the soil. Fir tuck, which exists mainly on exposed headlands, is formed when strong winter winds carry ice crystals that work like a sandblaster to kill the unprotected growing tips of the trees.

> **Think About It:**
> *Compare the list of common mammals of Newfoundland forests to those of Labrador forests. How would you account for the differences?*

Introducing New Species into Ecosystems

Imagine!
The boreal forest is the largest, coldest and slowest growing forest biome in the world.

Moose are one of the most important mammals of the forested areas of this province. Though we have large populations in Newfoundland now, they came from only six animals introduced to the island in two groups: a pair in 1878 and four more in 1904. Even snowshoe hare, which are now a very strong part of our environment and heritage, were introduced in the 19th century. With both these introductions, we were extremely lucky. They were brought into Newfoundland when no one was aware of the ecological disasters that have resulted from some of the introductions of animals into new territory. Perhaps the most well known of these was the introduction of rabbits to Australia where there were no natural predators.

All introductions of non-native species are potentially dangerous. It is practically impossible to predict how the new species will interact with and change the existing relationships in an ecosystem. For example, during the early 1960s, our government was considering bringing white-tailed deer to Newfoundland. It seemed like a good idea. After all, they did quite well in nearby Nova Scotia and it would provide another big game animal for hunting. At the time, a few people thought that it was risky and potentially damaging to the environment, but they could not predict what the specific damage might be. Several years later, we discovered that white-tailed deer carry a brain worm that causes them little harm. However, when that worm was experimentally injected into moose and caribou it proved fatal to both. It was concluded that, had we introduced white-tailed deer, we may have destroyed both our moose and caribou populations.

Red squirrel

Analysis:

10. *Research one other animal that has been introduced to Newfoundland.*

11. *Do you think the introduction was helpful or harmful? To whom?*

12. *Are there situations where the introduction of a species might be worth the risk? Explain.*

Boreal forest: northern coniferous forest.

Tundra: a treeless arctic area
usually underlain by permafrost.

Barrens

Few Canadians are as familiar with the barrens as Newfoundlanders and Labradorians. Anyone who has
picked partridgeberries has tramped over this magnifi-
cently wild, open country. We know them as barrens
mainly because they are treeless, not that they are barren
of life. Perhaps the barrens are best known as the home
of the caribou and the ptarmigan or 'partridge', two
animals wonderfully adapted to this windswept landscape.

Most of the barrens of Newfoundland and those of
southern Labrador are part of the **boreal forest** biome.
In fact, many of these barrens were once forested but
have become treeless as a result of repeated fires. Forests
did not regrow because fires eliminated much of the tree
seed stock, allowing shrubs to invade and dominate the
landscape. The barrens of the Long Range Mountains
and those of northern Labrador above the treeline are
part of the **tundra** biome.

The plants and animals that live in the barrens are
well adapted to cold, wind, and poor, generally acidic
soil. Most of the vegetation is composed of grasses, lichens, and shrubs like blueberries and sheep laurel.
These shrubs are part of a group of plants known as
heaths, hence the name *'heathland'* as the barrens are called

in other parts of the world. Few plants rise more than a foot or so above the ground and they grow slowly. The main animals of the barrens include arctic hare, arctic fox, lemmings, snowy owls, rough-legged hawks, horned lark and savannah sparrows, to name a few.

A Closer Look: Mina Hubbard's journal

Walking back along the point we found it cut by caribou trails, and everywhere the moss was torn and trampled in a way that indicated the presence there of many of the animals but a short time since. Yet it did not occur to me that we might possibly be on the outskirts of the march of the migrating caribou. Ptarmigan were there in numbers, and flew up along our way. We passed a number of stags, with antlers so immense that I wondered how they could possibly carry them. Beyond the lower slope of the hill seemed to be a solid mass of caribou, while its steeper part was dotted over with many feeding on the luxuriant moss.

Those lying along the bank got up at sight of us, and withdrew towards the great herd in rather leisurely manner, stopping now and then to watch us curiously. When the herd was reached, and the alarm given, the stags lined themselves up in the front rank and stood facing us, with heads high and a rather defiant air. It was a magnificent sight. They were in summer garb of pretty brown, shading to light grey and white on the under parts. The horns were in velvet, and those of the stags seemed as if they must surely weigh down the heads on which they rested. It was a mixed company, for male and female were already heading together. I started towards the herd, Kodak in hand, accompanied by George, while the others remained at the shore. The splendid creatures seemed to grow taller as we approached, and when we were within two hundred and fifty yards of them their defiance took definite form, and with determined step they came toward us.

The sight of that advancing army under such leadership was decidedly impressive, recalling vivid mental pictures made by tale of the stampeding wild cattle in the west. It made one feel like getting back to the canoe, and that is what we did.

Mina Hubbard
A Woman's Way Through Unknown Labrador

Mina Hubbard, a woman from a small Ontario town, was 35 years old when she decided to cross the uncharted Labrador wilderness from North West River to Ungava Bay in 1905. No other white traveller had managed to make this journey before her, including her husband who died during his attempt some years earlier.

Population Change on the Barrens

Ptarmigan

Population density: the number of organisms per unit area.

A population of animals is a group of the same species living close enough together to allow them to breed with one another. Breeding can occur between two animals from separate populations, but it is less likely than breeding within a population.

Consider the caribou of the Northern Peninsula and those of the Avalon Peninsula. Clearly, these are two separate populations. The distance between them is too great to allow interbreeding. Compare this with two caribou herds living closer together. It would be much more difficult to decide if they were acting as one population or two. It all depends on the amount of mixing that occurs between them.

Why is it important to know if the caribou are one population or not? Just imagine your job as a biologist if you were trying to determine how many animals were in a population and how fast the population was changing. You might be dealing with two populations that are changing at very different rates. One might be growing while the other is shrinking. If your job required you to set a harvest quota for these herds and you thought the caribou belonged to one population, not two, you might set the quota far too high for the dwindling population and exterminate it.

This same principle applies to all fish and wildlife populations. Knowing the size and distribution of populations is essential for their proper management and use. Imagine the difficulties in conducting a census of marine mammals like seals or cod fish!

If you were studying another animal of the barren-grounds, ptarmigan, for example, or even if you were hunting them, you would be interested in **population density**. This simply refers to the number of animals of a population that exists in an area of known size. Therefore, we refer to numbers like 10 ptarmigan per hectare or 200 ptarmigan per square kilometre. If the ptarmigan population density in an area were high, and you were a hunter, your chances of getting a meal or two would also be high.

How might the population density of an area be determined? Earlier in this course you learned about determining population densities of plants in a bog using quadrats. Plants don't move about like animals, so determining their density is easy compared to figuring out animal population density. This is one of the challenges for wildlife and fishery biologists.

Think About It:

If your job were to determine how many field mice (meadow voles) existed in an area, how might you do it?

Imagine!
The use of aerial surveys to count animal populations has become very costly. In 1992, it cost roughly $2,000 to count the moose in a one-kilometre square block of forested land. Many blocks of this size need to be counted in order to estimate the number of animals in one geographic area.

One method used to measure the number of ptarmigan in an area is the *drive or sweep count*. It is a very simple method, but requires many people. An area is chosen and its area is calculated. Then the people line up one next to the other, spaced a few metres apart. They proceed across the area from one side to the other, making sufficient noise to flush any birds that might be in the area. As the birds take flight they are counted. After the area has been covered, a density is calculated based on the size of the area walked.

Another more frequently used method involves only one or a few people. Each person walks a straight line transect across an area. Whenever a bird is flushed, the distance from the person to the take-off point is measured. This is called the flushing distance. After several transects are walked and a number of birds are flushed, the average flushing distance is calculated. This is then used as an indicator of the area covered by the person or the effective width of the strip. Based on this width, a density can then be calculated.

When animals can be seen from the air, one way to determine population density is to count them. This method is used for large mammals like moose, caribou and seals and also for some birds like those which nest in large colonies.

Determining the actual number of animals in an area is called measuring **absolute abundance**. This is often very difficult, and the amount and type of work required is often very expensive. The important thing

Aerial survey of caribou

Absolute abundance: an estimate of the actual number of organisms in an area.

Trend: a change in population, usually either an increase, a decrease or a tendency towards stability.

Relative abundance: an estimate of the number of organisms in an area at one time compared to another time, or the number of organisms in one area compared to another.

Winter ptarmigan

Limiting factor: an aspect of the environment that limits the size of a population.

Carrying capacity: the maximum number of a species that an area will support for a sustained period.

with many populations is not the precise number of animals in an area but whether they are increasing or decreasing. These **trends** or measures of **relative abundance** indicate how a population is changing and can help to determine if any action is necessary to control the change. To measure relative abundance, ecologists get some indication of how many animals are using an area at one point in time. They follow this with a later count, perhaps during a different season or in the following year. One example of this measure of abundance is the Christmas bird count, in which many people throughout North America participate each year (see page 74). To get a measure of relative abundance, ecologists don't have to count the animals themselves; instead, they can count the signs of animals or their tracks. This technique is often used in winter for animals that are hard to see.

Population density can change rapidly within a year, and more slowly over several years. For example, examine Figure 2.5, which shows a typical ptarmigan population over one year. Since each pair of ptarmigan might hatch an average of seven chicks each, the rapid growth in the early summer population should be no surprise. Chicks are most vulnerable soon after hatching since many are not fit enough to find adequate food, escape from predators, or avoid disease. Cold, wet weather also takes its toll on young birds. During late summer and early fall, young birds are stronger, can fly, and have figured out many survival tricks. During winter, food becomes scarce and weather conditions are harsh, so birds continue to die until the spring, when there is roughly the same number that were available to breed in the previous spring.

If this pattern were the same from one year to the next, the population in an area would remain roughly the same. However, changing conditions in the environment can set limits on the size of a population. The amount of food, water, shelter and space that is available are some of these **limiting factors**. Population size will also be affected by predators, disease, climate, pollution, and hunting. When one or more of these factors exceeds the limits of tolerance for the animal or plant, that factor will determine how many of the species will live in an area. This number is known as an area's **carrying capacity**. It is the maximum number of animals that an area will support for a sustained period. It can vary throughout the year and from year to year.

Too much of a limiting factor can be as bad as too little. For example, one of the limiting factors affecting plant growth can be soil acidity. Plant growth and reproduction will only occur if the acidity is within the required range for that particular plant.

As limiting factors change, they cause population levels to change as well. Wild animal populations are very rarely stable for long periods. For example, ptarmigan populations change dramatically, going from high to low and back to high about every ten years. This cycle is well documented now, but it is still unclear what the limiting factors are which cause the change.

Imagine!
Ptarmigan are found predominantly in the most northerly parts of Canada. Newfoundland populations are among the most southerly.

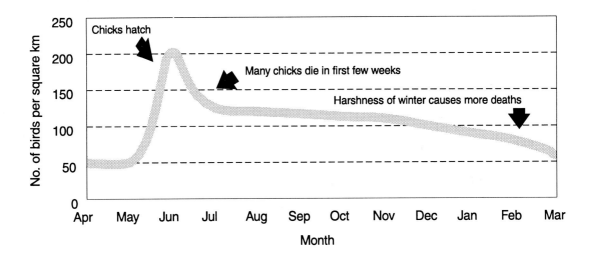

Figure 2.5 - Annual cycle of ptarmigan population.

Analysis:

13. *Describe two methods that can be employed to determine ptarmigan populations.*

14. *Why is it not possible to determine the 'absolute abundance' of moose in Newfoundland?*

15. *What human activities might affect the size of a caribou population? Explain your answer.*

A Closer Look: Christmas bird count

On Christmas Day in the year 1900, a group of bird enthusiasts from 25 different locations throughout the northeast of North America began a tradition—counting birds! Their goal was to find out the numbers and species of wintering birds, and where they were distributed. Since then, this has become an annual event. The number of count sites has increased from 25 to more than 1600, while the number of participants has grown from 27 to well over 42,000.

The Christmas bird count is the world's biggest birding event of the year. People participate in every Canadian province, every American state, Bermuda, many tropical American countries and numerous West Indian and Hawaiian islands. These groups all contribute to getting the most complete bird count possible on a given day, around Christmas day each year.

Each group is given a circular 'count area' of about 450 square kilometres. Within that 25 km-diameter circle, organizers attempt to place as many bird counters as possible. People are grouped into 'parties', each having a party leader who is familiar with their section of the circle. The count begins at midnight and continues for 24 hours.

The National Audubon Society sponsors and supervises the Christmas bird count. The Society collects the count information from all participating groups

Evening grosbeak

around the world, and then publishes the data each year. The annual count provides scientifically useful information on what species of birds are where, and in what numbers. But just as importantly, it is a social and competitive event that many bird enthusiasts look forward to all year.

Newfoundland participates in the count from a number of different locations—Terra Nova National Park, Corner Brook, Stephenville, Conception Bay South, Ferryland, Cape Race, St. Mary's, and St. John's. If you're near any of these locations around Christmastime, get more details on their bird count. Contact either one of the national parks, or ask for a representative from the Natural History Society in any of these areas. Who knows, you might be the one to make the first sighting of a rare and unusual bird!

The Ecoregions of Newfoundland and Labrador

What is an Ecoregion?

An *ecoregion* is a defined area where living organisms share the same basic living conditions. Like specific biomes (e.g. tropical rain forest), ecoregions are determined largely by climate. While biomes reflect global differences in climate, ecoregions reflect local or regional differences. In areas where climate and landscape are similar, the same communities of plants and animals will occur. Since climate is invisible, the plant communities are the primary indicators of ecoregions.

An ecosystem is more an idea than a place. In contrast, an ecoregion is a place. Ecologists can draw lines on a map showing the distinct ecoregions of an area (although some argue that we should not become too sure about drawing lines on dynamic environments). Newfoundland has nine ecoregions and Labrador has ten.

Ecoregions are described under five headings: *location, physical features, climate, vegetation* and *fauna*.

Ecoregions and Environmental Management

It is important for resource managers to understand ecoregions when they make certain environmental decisions. For example, in Unit 2 you will be looking at protected areas such as parks and wilderness areas. By protecting a certain portion of each ecoregion, we can ensure that a representation of its unique plants, animals and physical features are preserved for the future—for education, research and recreation.

The plant and animal communities in each ecoregion will respond differently to disturbances such as fire or logging. Before an area is changed, either accidentally

NEWFOUNDLAND ECOREGIONS
1. Western Nfld. Forest
2. Central Nfld. Forest
3. North Shore Forest
4. Northern Peninsula Forest
5. Avalon Forest
6. Maritime Barrens
7. Hyper-oceanic Barrens
8. Long Range Barrens
9. Strait of Belle Isle Barrens

LABRADOR ECOREGIONS
1. Low Arctic Tundra
2. Low Arctic-Alpine
3. High Subarctic Tundra
4. Coastal Barrens
5. Mid Subarctic Forest
6. High Boreal Forest
7. Mid Boreal Forest
8. Low Subarctic Forest
9. String Bog
10. Forteau Barrens

or intentionally, it is important to know what to expect after the change. For example, in ecoregion 2, the central Newfoundland forest, black spruce is the dominant tree in areas that were once burned. In some poorly drained areas, rich peatlands, called forested fens, develop. Logging in these areas should be avoided since heavy equipment will destroy the sensitive soils.

Newfoundland Ecoregions

Ecoregion 1: Western Newfoundland Forest

Hilly to mountainous. Most favourable climate on the island for plant growth. Warm summers, cold winters. Rainfall: 1000 to 1200 mm. Snowfall: 2 to 4 m. Heavily forested with balsam fir. Black spruce in sites with poor drainage or soil. Bogs are common. Endangered pine marten (a small mammal) around Little Grand Lake.

Ecoregion 2: Central Newfoundland Forest

Gently rolling to hilly. Warmest summers and coldest winters on the island. Rainfall: 900 to 1300 mm. Snowfall: 3 to 3.5 m. Heavily forested with balsam fir, black spruce and white birch. Burned areas have large stands of black spruce and white birch. Endangered red pine occurs in small pockets. Bogs are common.

Ecoregion 3: North Shore Forest

Coastal, rolling to hilly. Driest soils on the island. Warm summers, cold winters. Rainfall: 900 to 1200 mm. Snowfall: 2.5 to 3.5 m. Mainly forested with balsam fir and black spruce. Barrens on coast. Many seabird colonies along the coast such as Little Fogo Island.

Ecoregion 4: Northern Peninsula Forest

Flat to hilly. Short cool summers, cold long winters. Rainfall: 900 to 1000 mm. Snowfall: 3 to 3.5 m. Mainly forested with balsam fir except in the west where there are many bogs. Also barrens along west coast. A number of seabird colonies along the coast, especially eider ducks.

Ecoregion 5: Avalon Forest

Low hills with steep sides. Cool summers, mild winters. Rainfall: 1400 to 1500 mm. Snowfall: 2 to 2.5 m. Heavily forested, mainly with balsam fir. Many lichens on trees, especially old man's beard. Bogs in valleys.

Ecoregion 6: Maritime Barrens

Small rolling hills. Cool summers, mild winters. Rainfall and snowfall combined: 1250 to 1600 mm. No permanent snow cover except in southern interior. Mostly low shrubs, such as blueberries and laurel, with trees only in valleys. Bogs are common. Many major seabird colonies along the coast such as Witless Bay. Major caribou herds.

Ecoregion 7: Eastern Hyper-Oceanic (high coastal) Barrens

Flat to rolling coastal headlands. Very cool summers, mild winters. Rainfall: 1250 to 1450 mm. Snowfall: 2 to 2.5 m. No forest cover except for stunted conifer trees (tuckamore). Mainly low shrubs and mosses. Some major seabird colonies such as Cape St. Mary's.

Ecoregion 8: Long Range Mountains

Mountainous highlands. Cool summers, cold winters. Rainfall: 1000 to 1600 mm. Snowfall completely covers the land until late spring, can be over 5 m. Low shrubs and tuckamore in many areas. Bogs are common. Major caribou herds. Arctic hare found mainly in this ecoregion.

Ecoregion 9: Strait of Belle Isle Barrens

Rocky, flat, coastal plain in the west, hilly in the east. Cool short summers, long cold winters. Rainfall: 760 to 900 mm. Snowfall: 2.5 to 3.0 m. Longest period of snow cover on the island. No forest. Tuckamore is common.

Labrador Ecoregions

Ecoregion 1: Low Arctic Tundra - Cape Chidley

Flat coastal plain in the north, steep hills in the south. Short cool summers, very long cold winters. Rainfall: 500 to 600 mm. Snowfall: less than 3 m. Landscape is treeless, dominated by rock and open ground with patches of mosses and lichens.

Ecoregion 2: Low Arctic Alpine - Torngat

Mountains with deep fjords. Short cool summers, long, very cold winters. Rainfall: 500 to 700 mm. Snowfall: 3 m. Treeless with some meadows. Home of the Torngat Mountain caribou herd. Some polar bears make dens along the coast.

Ecoregion 3: High Subarctic Tundra - Kingurutik/Fraser

High plateaus cut by long fjords. Short cool summers, long, very cold winters. Rainfall: 700 to 1000 mm. Snowfall: 3 to 4 m. Tundra on highlands and forested valleys. Northern treeline at Napoktok Bay. Very important area for caribou (George River herd).

Ecoregion 4: Coastal Barrens - Okak/Battle Harbour

Coastal with exposed headlands, numerous islands and inlets. Cool summers, cold winters. Rainfall: 1000 to 1300 mm. Snowfall: 3 to 4 m. Ground-hugging shrubs, much exposed rock, forest in sheltered valleys. Major seabird colonies on islands. Several species of seal give birth along the coast and on islands.

Ecoregion 5: Mid Subarctic Forest - Michikamau

Flat to rolling plateau. Cool short summers, long cold winters. Rainfall: 900 to 100 mm. Snowfall: 3.5 to 4.5 m. Open lichen woodlands dominated by black spruce.

Ecoregion 6: High Boreal Forest - Lake Melville

Rolling uplands with flat coastal plain. Warmer summers and shorter winters than surrounding ecoregions. Rainfall: 800 to 1000 mm. Snowfall: 4 m. Very productive forests dominated by balsam fir, white birch and aspen. Bogs along the coastal plain. Many birds normally found further south occur only in this ecoregion in Labrador.

Ecoregion 7: Mid Boreal Forest - Paradise River

Small rolling hills. Cool summers, short cold winters. Rainfall: 1000 to 1200 mm. Snowfall: 4 to 5 m. Black spruce and balsam fir forests. Bogs in valleys.

Ecoregion 8: Low Subarctic Forest - Mecatina River

Rolling hills with broad river valleys. Warm summers, cold winters. Rainfall: 1000 to 1300 mm. Snowfall: 3.5 to more than 5 m. Mainly open black spruce forest. Large bogs throughout.

Ecoregion 9: String Bog - Eagle River Plateau

Flat to rolling plateau. Cool summers, cold winters. Rainfall: 1000 to 1200 mm. Snowfall: 5 m. Extensive bogs with much open water and stands of scrubby black spruce on bog hummocks. Some forested areas. Important area for waterfowl.

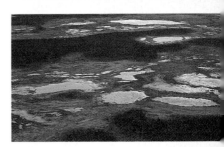

Ecoregion 10: Forteau Barrens

Flat-topped hills. Cool summers, mildly cold winters. Rainfall: 1000 to 1250 mm. Snowfall: 3.5 to 4.5 m. Scrubby spruce barrens with some bogs. Forests in river valleys.

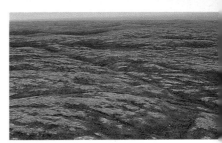

Activity: How It Was, How It Is... How Will It Be?

Arrange to interview an older individual from your community. Ask them questions about the way they remember the local environment when they were younger. Try to get detailed information about specific areas, maybe a pond, or river or a marsh where they once picked bakeapples.

Then, visit the places you talked about in your interview. Describe any changes that have taken place, both positive and negative. What caused those changes? Record your reactions to your findings. Next, try to find out what is going to happen to these areas in the future. Are there plans for developments of any kind? What would you like to see happen? How will it be in the future?

Analysis:

16. With an increased concern for our local, regional and global environments, do you think there will be an increased demand for individuals trained in environmental science? What type of jobs might be needed?

17. How would you react to a person who argued that people should stop disturbing natural ecosystems, harvesting trees, and fishing in the Northern Atlantic because it is damaging? What would you say to that person?

18. Having finished the first unit of this environmental science course now, have your views, thoughts or feelings about the environment changed in any way? If so, how?

19. You have just studied and experienced some of the most common ecosystems in Newfoundland and Labrador. Make a table showing some of the uses and abuses of these ecosystems under the following headings: recreation and commercial use.

20. What is your favourite ecosystem and why? If you were given five acres of your favourite ecosystem what would you do with it?

21. "Everything depends on everything else." In your journal, try to bring this statement to life with an essay, poem, words to a song or a drawing.

Last Thoughts

Having first viewed the whole earth and then the ecosystems and ecoregions of Newfoundland and Labrador, you have built a foundation here for what is to follow in this course. We will return to these ideas frequently as we look at the impact of our use of natural resources in this province and on our planet.

Always bear in mind that our culture, and our social, economic and political activity can never be separated from the ecosystems upon which they rely. Our culture has grown out of a heavy reliance on our natural resources, and in a sense, our ecosystems have created the kind of people we are and what we do.

As you proceed, it is important not just to build your knowledge and skills, but to feel for and care about the world around you. In order to form valid opinions on many of the topics to follow, you will need to understand the issues involved and to develop or clarify your own values. These will reflect what is important to you.

There is always a temptation to see issues in 'black and white', as right or wrong, but there are almost always other valid perspectives. Good decision-making usually requires that we consider the different points of view until it becomes clear that the weight of fact and opinion favours a particular decision. The development of this skill is an important part of Units 2 and 3.

Our plans to manage the Earth are founded on the belief that ignorance is a problem that can be solved with science and technology...But there is much to learn about the Earth, and there are many good reasons to believe that its complexities are permanently beyond our comprehension. Thus the salient point is our ignorance, not our knowledge.

Chris Maser

NORTHERN COD

MUNICIPAL WASTE MANAGEMENT

LOW-LEVEL FLYING IN LABRADOR

UNIT 2
Local Environmental Issues

MOOSE AND CARIBOU MANAGEMENT

WILDERNESS AREAS—BAY DU NORD

NEWFOUNDLAND PINE MARTEN

FOREST MANAGEMENT IN THE 1990s

PULP AND PAPER INDUSTRY

MINERAL EXPLORATION AND DEVELOPMENT

AQUACULTURE

Unit 2

These next ten chapters are about problems in our own backyard in Newfoundland and Labrador. The issues range from waste management to endangered species; from managing our forest, wildlife and mineral resources to understanding the challenges ahead for our troubled fishing industry.

These are not simple problems, nor are there simple solutions. You will encounter a variety of perspectives and arguments here. Be prepared to ask questions, and be equally prepared for more than one answer.

This unit provides a good opportunity to build your base of knowledge through investigation of the concepts and perspectives surrounding each of these issues. You will have heard something about many of these issues before, most likely through the media. However, in order to keep these issues topical and relevant, it will be important for you to add today's newspaper clippings, magazine articles, and other information sources to the material you find here.

Unit 2 is a slice of the real world—our real world in Newfoundland and Labrador. You will encounter actual places and people in the province. You will see that the problems facing our environment are not just 'out there somewhere.' They are here, too. And you can be part of the solution.

What you will learn:

- the basic scientific concepts and factual information underlying local environmental issues;
- the problems associated with these local issues;
- the gathering, interpretation and evaluation of information relating to issues;
- an analysis for bias in points of view;
- the human element in environmental issues;
- an exploration of your own opinions and the information and values on which they are based;
- the relationships among science, technology and society in environmental issues;
- a sense of your personal responsibility and our collective responsibility to work towards solutions to these issues.

CHAPTER 3
Northern Cod

There's lots of fish in Bonavist' Harbour
Lots of fish right in around here...
Anonymous

What you will learn:

- the importance of northern cod to the history, culture, social structure and economy of the province;
- what a fish stock is and the methods used for assessing cod stocks;
- the difficulties associated with measuring fish populations;
- the distribution and spawning grounds for the northern cod stock;
- the differences between the inshore fishery and the offshore fishery;
- the various other impacts and influences on the northern cod resource;
- the possible causes for changes in the status of the northern cod stock;
- the role of the Department of Fisheries and Oceans in managing the northern cod stock;
- the contributions of science and technology in the management of northern cod;
- some of the implications of the 1992 northern cod moratorium on the province.

Imagine!

The biggest codfish recorded for Newfoundland was caught in 1949 off Spare Harbour, Labrador by George Earle from Carbonear. The fish was 68 kg and 1.8 metres long. Imagine Mr. Earle's shock as he hauled this giant out of the sea, caught on his jigger, by the tail!
In 1895, an even larger cod was caught on a linetrawl off Massachussetts, weighing 96 kg and measuring nearly 2 metres long.

Introduction

When we talk about fish in this province, it is generally understood that we are talking about codfish—the most important of our fish resources.

In 1497, John Cabot returned to England proclaiming that cod were so plentiful off our coasts that they sometimes stopped the progress of his ships. Whether or not his reports were exaggerated, it is a fact that, since that time, the cod fishery has shaped our identity as a people. We live in communities that have depended on this resource for more than 400 years. It has contributed not only to our settlement patterns, but also to our economy, our lifestyle, our culture, our diet and even the songs we sing.

Atlantic cod are found on both sides of the Atlantic Ocean. In the northwest Atlantic, their range extends north to Greenland and south to Cape Hatteras, North Carolina. However, cod are most abundant along Canada's Atlantic continental shelf—the relatively shallow waters separating the land from the ocean depths (see Figure 3.1). These banks provide an ideal habitat for fish. Currents, storms and other causes of movement of the ocean layers create the right kind of conditions for stirring up nutrients and for growing plankton, an important food for fish.

Northern Cod
The Heart of the Fishery

The most important population of Atlantic cod in the Newfoundland region is the 'northern cod stock.' This group of fish is found off eastern Newfoundland and Labrador and has historically made up three-quarters of all fish landed in the province. More than 60% of our fishermen and nearly 70% of our plant workers have depended on this resource for their livelihood. In 1991, the northern cod fishery was worth $700 million to this province.

In 1992, 19,000 people earned a full-time living directly from the northern cod fishery. On Friday July 3,

1992, those same people were out of work. The industry was stopped dead when a two-year **fisheries moratorium** was announced for northern cod. This resource had been drastically reduced, and a complete halt to the fishery was considered necessary if the fish stocks were to recover.

How did we arrive at such a crisis in the fishery? Who is responsible? How is the province affected? Can the problems be fixed? These are questions without easy answers, but in the pages that follow we will explore some of the pieces of the northern cod puzzle in an effort to understand.

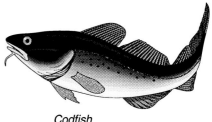

Codfish

Analysis:

1. *What conditions along the Grand Banks help create an ideal habitat for fish?*

2. *What event occurred on July 3, 1992? Why was it necessary to take this action?*

Northern Cod Science—A Matter of Scale

We have been harvesting the codfish for four centuries, but there is still a great deal we don't know about this species. We can't see it as an entire population, so it's very difficult to measure, and nearly impossible to predict. Fisheries managers within the Department of Fisheries and Oceans (DFO) require good scientific data in order to make decisions such as how many fish can be safely caught without jeopardizing the health of the population, how they should be fished, and from where.

Fish of a Feather 'Stock' Together

Within the distribution area for cod in Atlantic Canada, fisheries scientists have identified more than a dozen different **fish stocks**. The northern cod stock includes fish in at least four spawning areas, including the Hamilton Bank, the Belle Isle Bank, the Funk Island Bank, and the northern Grand Bank. There is also evidence that 'bay stocks' exist which spawn in inshore areas (see Figure 3.1).

In Chapter 2 (p. 70), you learned about the importance of recognizing certain populations of the same animals as being distinct from each other. The same applies

Fisheries moratorium: a cessation of fishing of a particular species or population of a species.

Apart from sound science, there is no other acceptable source of appropriate management advice...Good science is still the essential key to our problem, but until such time as we have absolutely perfected our techniques, we must remain willing to submit our uncertain data to every reasonable test to confirm or reject it.
Dr. Leslie Harris

Fish stock: a group of fish that has a common genetic make-up, inhabits and reproduces in a particular region and maintains a similar migration pattern.

Imagine!

Although caplin is their favourite food, codfish will eat almost anything they can find. In 1871, a fisherman in St. John's was splitting a large codfish and noticed that the fish had a full stomach which contained shells, fish bones, crab parts, and a gold wedding ring with the initials of its owner and a date inside the band. The owner had been a passenger aboard the steamship "Anglo Saxon" which was lost off Chance Cove ten years before!

From Jensen's *The Cod*

A postage stamp was issued by the Newfoundland Government in 1932 and remained in circulation until 1937. The inscription reads: "Codfish Newfoundland Currency"

From "Fisheries Calendar," DFO, 1992.

I used to go out onto the porch and see the boat coming, right flat on the water, loaded full of fish. I often said, 'now, thanks be to God, ain't that a lovely thing to see.'

Mary Jo O'Keefe

to fish stocks. Heavy commercial harvesting makes this distinction between stocks even more vital, since it is critical that we don't seriously overfish any one group. When deciding how many fish can be harvested, fish managers must consider each stock separately.

In 1977, Canada's 200-mile offshore economic zone was established along with other fishing zones, or divisions, for all commercially fished species within the northwest Atlantic region. The northern cod stock is commonly referred to as all the fish within divisions 2J, 3K, and 3L (see Figure 3.1).

What Makes a Cod Want to Move?

Scientists suspect that there are three main things that motivate cod to move—feeding, spawning, and climate. There are two periods during the year when the fish are most heavily concentrated. During the winter months, most cod can be found spawning on the offshore banks (see Figure 3.1). In late June and early July, some of the fish move inshore, perhaps in pursuit of caplin.

Ocean climate is another factor influencing the movement of cod. Water temperature, currents and salt content all affect fish. For example, cod generally remain in waters where the temperature is between 0° C and 4° C. If the water temperature increases or decreases beyond that range, the fish will move elsewhere.

Chance of Survival—One in a Million

Once a female cod reaches sexual maturity, she lays an average of one million eggs. This may seem like an enormous number, but alas, life as a young codfish is no picnic. Predators, tides, currents, temperature changes and pollution all contribute to a mortality rate that will see only about one in a million of those eggs reach sexual

Analysis:

3. What are two factors that influence cod migration?

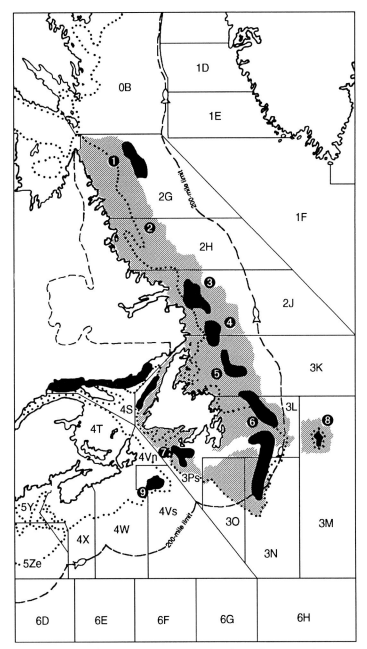

All hands would have to scrabble... you'd see three-year-olds goin' off with a fish almost as big as theirselves. It was hard work, but good work.

Bessie Flynn

LEGEND

❶ Saglek Bank
❷ Labrador Shelf
❸ Hamilton Bank
❹ Belle Isle Bank
❺ Funk Island Bank
❻ Grand Bank
❼ St. Pierre Bank
❽ Flemish Cap
❾ Banquereau Bank

· · · Continental Shelf
– – 200-mile limit
▮ Spawning areas
▨ Distribution of Cod Stocks

Figure 3.1 - Fishing zones, and distribution of cod stocks

89

maturity (in approximately six or seven years). If an average of two offspring from each female can live to be seven years old, the stock should remain fairly stable.

Pressures on Northern Cod

Like most marine species, the northern cod faces natural pressures such as predation, climatic change, and death from other causes. However, it runs a far greater risk of meeting its fate at the hands of the inshore or offshore fishery, or by foreign trawlers. Until the summer of 1992, the offshore and foreign fisheries were allowed a certain **quota** of the **Total Allowable Catch (TAC)** set for each year. The inshore fishery was also permitted a certain **allowance** of the overall quota. Now that the northern cod stock has been severely overfished, each of these groups is attempting to pin the blame elsewhere, away from themselves. See the margin for a somewhat comical look at this finger-pointing syndrome.

Is it possible to place sole blame on any one of these groups? Or does the present situation reflect more widespread abuse of the resource? Let's look first at those who fish for northern cod.

The Inshore and Offshore Fisheries

In the years leading up to the 1992 northern cod moratorium, inshore fishermen attempted to make known their concerns about the smaller and fewer fish they were catching. It was taking more gear, fuel and time to catch the same amount as they once did. This represented a decrease in what is known as the 'catch per unit effort,' where less is caught with the same amount of effort.

They couldn't help but turn their frustrated and angry gaze seaward, to the huge company-owned fleets that dragged the bottom for northern cod—in the winter while the fish were spawning.

Meanwhile, the offshore trawlermen claimed that codtraps were taking large numbers of young fish that may not yet have reached spawning age. See Figure 3.2 for a historic look at the landings of northern cod by the **inshore** and the **offshore** fishery.

Who's to Blame?

The offshore trawler captains do not think there is a problem, except for seals that eat the inshore cod; the inshore fishermen believe the offshore fishermen are the problem; ...the foreign experts say it is Canadian mismanagement; the provincial government blames the federal government; an independent study suggests a serious resource problem exists, one that stems partly from inaccurate DFO estimates of the cod stocks for the past 13 years; the scientists who made the estimates deny there is a resource problem; and a woman at the Petro-Canada station in Goobies believes it is all God's will.

Don Gillmor, *Equinox*, No. 52, July/August 1990

Quota: amount of catch allowed the different sectors of the fishery.

Total Allowable Catch (TAC): total catch allowed from a particular stock over a given period.

Allowance: amount of catch allowed the inshore sector; a portion of the overall quota.

Inshore fishery: a shore-based fishery conducted mainly during the summer months using hand lines, gill nets, and/or cod traps.

Offshore fishery: a fishery conducted from large trawlers that remain at sea for weeks at a time.

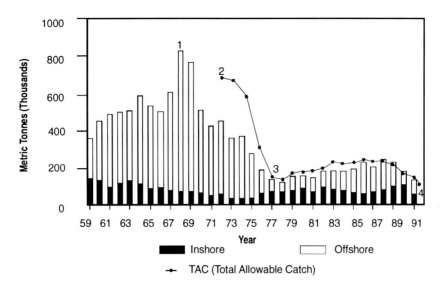

Figure 3.2 - Northern cod landings and Total Allowable Catch (TAC) in NAFO divisions 2J 3K 3L, 1959-1991. (See margin notes.)

The Foreign Fishery

Fish are a lot less interested in lines on a map than we are. And unfortunately for us, the outer edges of the Grand Banks, commonly referred to as the 'nose' (3L) and 'tail' (3NO) of the Grand Banks, lie outside of the 200-mile limit and therefore outside of Canadian waters (see Fig. 3.1). So in studying pressures on the northern cod stock, it is also necessary to look at international fishing both inside and outside the 200-mile limit in 2J3K3L as well as within the bordering zones of 2GH and 3NO.

There is a quota set each year for foreign vessels, agreed upon by Canada, the European Economic Community (EEC), and other foreign countries such as Korea, Cuba and Japan. Between 1985 and 1988, European nations (Germany, Portugal, France, Spain, and U.K.) were given a quota of 35,000 metric tonnes for northern cod. They reported to have landed 165,000 metric tonnes—almost five times the amount they were allocated. The Northwest Atlantic fisheries Organization (NAFO) has jurisdiction for managing, regulating and enforcing the fishing zones outside the 200-mile limit. However, NAFO's role is often ineffective, as demonstrated in 1991 when it declared a moratorium on fishing cod on the

Refer to Figure 3.2

1. This peak in 1968 showed landings exceeding 800,000 metric tonnes. This marked the culmination of about a decade of significant technological innovation in the fishery. Note the resulting crash in the years that followed—the lowest levels ever recorded.

2. In 1972 the International Commission for Northwest Atlantic Fisheries (ICNAF) set the first Total Allowable Catch (TAC) in an attempt to regulate and manage the stocks beyond what was then only a 12-mile coastal limit.

3. In 1977 Canada negotiated declaration of the 200-mile economic zone, giving us full control to manage stocks within that zone. In the following year, the Northwest Atlantic Fisheries Organization (NAFO) was established to replace ICNAF as the regulatory and management agency for fishing zones beyond the 200-mile limit.

4. In July 1992 a northern cod moratorium was declared. The Canadian Atlantic Fisheries Science Advisory Committee (CAFSAC) advised NAFO that "the abundance of cod had declined drastically in the previous 12 to 18 months."

'nose' of the Grand Banks. EEC countries objected to this and caught an estimated 50,000 tonnes of cod despite the moratorium.

Foreign countries are also allowed quotas for other species besides cod in these areas, and in these catches, cod are often caught 'by accident' (see margin note on 'Bycatch'). To add to the problem, there have been accounts of under-reporting of up to 25% of the total catch, and dumping of under-sized fish at sea.

Analysis:

4. *Read the excerpt from the magazine article by Don Gillmor on page 90 of the text. Make a list of the possible causes of the decline in the cod populations. What do you believe is the cause or causes? What do you base your answer upon?*

5. *What political pressures do you think might exist in the allocation of quotas to members of NAFO (Northwest Atlantic Fisheries Organization)?*

6. *Look carefully at Figure 3.2. Describe what happened between the years 1968 and 1978. What were the reasons for the differences in catch levels?*

Managing the Northern Cod Stock

The federal Department of Fisheries and Oceans (DFO) has the responsibility of managing the northern cod stock. Their goal is to allow for the maximum *yield*, or catch, while maintaining the health of the stock biomass. DFO is responsible for setting the TAC and quotas, as well as the licensing of fishermen.

Stock Assessment—The Scientific Basis for Management

The estimation of how much fish is out there and the effects of fishing on the stocks is known as stock assessment. This is the most important data on which decisions are made regarding the management of the fisheries. A large percentage of all fisheries research undertaken by DFO in the Newfoundland region is in stock assessment and the provision of biological advice to fishery managers. The northern cod stock is assessed each year, and this information forms the basis for deciding how much fish can be caught for that year—Total Allowable Catch (TAC).

Bycatch

The issue of bycatch—catching fish you did not set out to catch —is an issue of real concern as it relates to northern cod. It is a problem not only with the foreign fishery, but also with our own offshore and inshore fisheries. In the past, foreign countries were allowed to keep 10% bycatch of cod when fishing for other species. However, if that country also had a cod quota, that percentage of bycatch was deducted from their cod quota. With the current ban on fishing for northern cod, there are greater efforts being made to minimize bycatch of cod by all fishing sectors.

In attempting to arrive at a population estimate, fisheries scientists must rely on a variety of data sources. They include observations made by scientists at sea on research vessels (RVs) and/or commercial fishing vessels, as well as information obtained from commercial harvesters regarding the catch and catching process (see margin).

Data from these sources is then used in what is called **Virtual Population Analysis (VPA)**. This involves counting the total number of fish caught from one particular **year-class** over a period of years until no more catches of that year-class are reported. This analysis is done for each year-class, which are added together to get an idea of the overall population. Added to these figures is an estimate of natural mortality (from predation and other natural causes). In other words, we only really have an idea of population sizes after any one year-class has been depleted.

In actual fact, scientists cannot rely solely on numbers. They must combine an estimate of population numbers with an estimate of population ages (and thus sizes). This leads to an approximate measure of the total weight of fish in a stock, which is known as the fish **biomass**. It is this number that is most important in terms of managing the commercial fishery.

Another important scientific measurement is that of the spawning success of a certain year-class of fish within a stock. Since 1991, DFO has conducted surveys to determine the distribution and numbers of eggs, larvae and young cod. From this data, along with studies of juvenile cod, an estimation is made of the survival of a certain year-class to the age when it is big enough to be caught in the commercial fishery (about three to five years old). This rate of survival is called the *recruitment rate*.

All of these methods provide vast amounts of information for fisheries management. Imperfect though it is, scientists use the data to estimate the population numbers and biomass for a given stock. With that information, fisheries managers then provide advice to the government of Canada on what the TAC should be for that year.

But science is not the only consideration in managing the fishery.

Data Sources for Stock Assessment

- **Catch Data**: the weight of fish caught by both the inshore and the offshore sectors, sorted according to fishing gear type, vessel type and month.
- **Catch Per Unit Effort (CPUE) Data**: the weight of fish landed given a specific level of effort (i.e. amount of time and gear-type).
- **Research Vessel (RV) Survey Data**: estimates of population numbers and biomass through systematic surveys of the stock area.
- **Commercial Sampling**: analysis of the age, length, weight and sex of fish landed by both the inshore and offshore sectors.

Virtual Population Analysis (VPA): a method used for estimating absolute population size of a species by looking back over time at the numbers of fish caught of a particular year-class, and adding to this an estimate of natural mortality.

Year-class: animals that hatch or are born in the same year and are therefore the same age.

Biomass: the weight of a stock, stock complex, or population.

We do not envy the politicians who must make such difficult choices. We do, however, insist that such choices be made... We also insist that quite apart from social and economic concerns, which understandably assume a dominant role, the Government of Canada has the unequivocal obligation of conserving one of the great living natural resources of the nation.

Northern Cod
Review Panel, 1990

Fish, Politics and People

Politicians also take into consideration the needs of people when they make their decisions, and these needs do not always dictate what is best for the resource, or even for people in the long run. Dr. Leslie Harris, who chaired a study team looking at the northern cod stock, concluded the following:

> *The most important thing we have to do is take politics out of the management decision-making process: that's very easy to say and difficult to do. But I think that if we're going to sustain the stocks, they have to be managed on the basis of objective scientific criteria and it means we have to eliminate decision-making on the basis of expediency.*

Activity: Age a cod—look in its ear

Age is a key factor in determining population numbers and thus in calculating population biomass. The older the fish, the larger and heavier it is. It is also important to determine approximately how many fish within a stock are from different 'year-classes'—fish that hatched in the same year and are therefore the same age.

Scientists are able to determine the age of a cod by counting the growth rings on a cross-section of its earbone, or otolith (see illustration to the right). How old would you estimate the fish who owned this otolith to have been when it was caught? Codfish generally reach sexual maturity at about six or seven years of age. Had this fish spawned before it was caught?

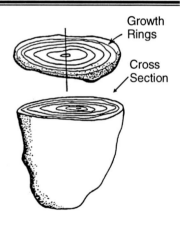

Growth
Rings

Cross
Section

Analysis:

7. Summarize the different information sources used by fisheries scientists to assess fish stocks.

8. What are some of the reasons for difficulties in estimating the cod population?

9. Since fisheries scientists rely heavily on catch data (i.e. amount of fish landed) to determine population biomass, how do you think they will assess the northern cod stock when no fish are being caught?

10. It is suggested that there has been too much fishing pressure on the northern cod resource. If you were in a position to bring about changes in the management of northern cod, what alternatives would you propose?

A Closer Look: fishing songs are different now

Newfoundland's culture, and specifically its folk song traditions, grew out of people's relationship to the sea. Hundreds of songs were written about fishing in a time when fish were plentiful...*I'se the bye that catches the fish; There's lots of fish in Bonavist' Harbour; Lukey he sailed up the shore to get some fish from Labrador; He fished in Indian Harbour where his father fished before; Let me fish off Cape St. Mary's;* and many more.

The following song was written by Jim Payne, a modern-day folksinger/songwriter. How do the content and tone of the lyrics portray a new era for Newfoundland's fishery?

Empty Nets
Get up in the morning at a quarter to four,
Try not to make much noise as you go
  through the door;
Jump in the boat, you can hear the gulls roar
At the start of a brand new day.
Fire up the engine, you're ready to go;
Head out the harbour, you don't want to be
  slow.
What's out there today? Well, you never
  know -
Just hope that it turns out OK.

Chorus:
'Cause it's empty nets, that's what he gets,
When you're out on the water no time for regrets.
Those empty nets, that's what he gets.
How's a poor fisherman to pay off his debts
When he goes out each morning to haul empty
  nets?

You can blame it on foreigners, blame it on
  feds,
Cast all the blame on each other instead.
But when all's said and done, it's still something I dread—To see Newfoundland give up
  the fishery.
What of our communities? Will they just die?
"Pack up your duds, give the mainland a try."
But I'm staying here, 'til someone tells me
  why I should put up with this misery.

Here's to the plant worker, toils on shore.
Waits for the fisherman to catch a few more.
Then pack it up for the grocery store
'Til it ends up on somebody's table.
How can they feed multitudes on fishes so
  small?
How can they feed families on no fish at all?
Get down on your knees for a miracle call;
We'll stick it for as long as we're able.

Here's luck to the fisherman, he'll need it I
  know,
As he bobs on the ocean, God bless his poor
  soul.
May good fortune follow wherever he goes,
To keep him from debtload and danger.
And wherever you live, no matter which bay,
May bankers and loan boards not stand in
  your way;
May you bring home a boatload each single
  day,
And to poverty be ever a stranger.

(Song by Jim Payne. Written for the Coalition for Fisheries Survival and Memorial University of Newfoundland's fisheries forum, June 1990.)

From the compact disc "Empty Nets" by Jim Payne
SingSong SS9192 © Jim Payne SOCAN

Empty Nets—Why?

We are kidding ourselves if we think there is any easy answer to the question of how and why the northern cod stock has reached the point of near destruction. In this chapter, we've learned about a number of factors influencing cod populations, but the reality is we've only scratched the surface of what needs to be understood.

The marine ecosystem is a complex web of interrelationships, made more complicated by the fact that we cannot see what we are dealing with. We don't really understand predator/prey relationships enough to know the impact of seals on cod; or the implications of fishing for caplin, the most important food species for cod. We don't really understand what kind of effects climate change, pollution and offshore development activity have on marine species. And there is a host of other things we don't understand. How can we presume to manage something we know so little of?

The reason is people. People are part of this equation. People need jobs, and people need to feel connected to the livelihood on which their culture and lifestyle is built. People from around the world continue to place increasing demands on the resources of the sea. And as the world population rises, so does its appetite. So we continue to fish, despite not knowing the impact we are having on this and other marine species.

Technology has been called 'a wonderful servant, a terrible master.' The reality of the northern cod problem is that our technology for learning about fish is decades behind our technology for catching fish. The decline of the northern cod stock may well have begun in the late 1960s when there were massive technological advances made in fishing gear, vessels, navigational and acoustic equipment, processing capabilities, and more.

Finally, there is the distinct possibility that the fate of the northern cod is caused by what biologist Garrett Hardin has called 'the tragedy of the commons'— too many people staking claims to the same one resource. It is common property, so there is little incentive for individuals and corporations to regulate their behaviour.

Offshore trawler

Fisheries Short Form— Organizations and Programs

CAFSAC - Canadian Atlantic Fisheries Scientific Advisory Committee

CPUE - Catch per Unit Effort

DFO - Department of Fisheries and Oceans

EEC - European Economic Community

FFAW - Fishermen, Food and Allied Workers

ICNAF - International Commission for the Northwest Atlantic Fisheries

NAFO - North Atlantic Fisheries Organization

TAC - Total Allowable Catch

VPA - Virtual Population Analysis

Think About It:

"The tragedy of the commons develops in this way. Picture a pasture open to all. It is to be expected that each herdsman will try to keep as many cattle as possible on the commons... The rational herdsman concludes that the only sensible course for him to pursue is to add another animal to his herd. And another; and another... But this is the conclusion reached by each and every rational herdsman sharing a commons. Therein is the tragedy. Each man is locked into a system that compels him to increase his herd without limit. Ruin is the destination toward which all men rush, each pursuing his own best interest in a society that believes in the freedom of the commons. Freedom in a commons brings ruin to all."

Garrett Hardin

In what ways is the northern cod issue a kind of "tragedy of the commons"?

Last Thoughts

July 1992 not only marked the date of the largest single layoff in Canadian history, but it also dealt a huge blow to the heart and soul of our entire province. Although it was almost unbelievable that the fish were gone, the people whose history was rooted in the fishery were told to stay home.

Perhaps the northern cod stock will come back again. But there is always a chance that it will not. As in any tragedy, we must ask ourselves an important question—are we capable of learning from this?

There is always an easy answer to every human problem—neat, plausible, and wrong.

H.L. Mencken

Students in Action – become informed

As an individual, one of the best actions you can take to assist the fisheries is to get involved and keep informed. Find out more about any of the following topics and others you can think of:

- Fishworkers' unions and associations
- Foreign overfishing
- Quota allocations
- Cod farming
- Northern cod science
- Relationships among fish, whales, seals and caplin
- Licensing
- Marketing fish
- Report of the Northern Cod Review Panel by Dr. Leslie Harris.

CHAPTER 4
Municipal Waste Management
A Growing Concern

Whatever befalls the earth, befalls the sons of the earth. Man did not weave the web of life; he is merely a strand in it. Whatever he does to the web, he does to himself. Continue to contaminate your bed and you will, one night, suffocate in your own waste.

Seattle, Chief of the Dwamish Indians, 1851

Introduction

Everything we do results in some kind of waste. Activity requires energy, which we get from food. We create waste when we produce food, package it, transport it to market, drive to the store to buy it, prepare it for our tables, and we create more when we eat it. This is a natural cycle. But modern society is now producing garbage at a rate that may be beyond the ability of natural systems to manage.

After the Second World War, North America experienced a great economic boom. The result was a complete change in patterns of production and consumption in our society. An increase in population, changes in lifestyle and a desire for disposable materials have led to a tremendous increase in the amount of waste produced (see Figure 4.1). Canadians are among the most wasteful people in the world. It is estimated that we throw away from 1.7 to 3.0 kilograms of garbage per person per day. In the area of St. John's, where 160,000 people use one disposal site, about 270,000 kilograms, or 75-100 dump truck loads of garbage are tossed out each day!

Types of Waste

Our society produces four types of wastes—industrial waste, agricultural waste, sewage waste and municipal waste. Municipal waste consists of all the materials that are thrown away from homes, businesses and institutions such as schools (see Figure 4.2).

All waste materials can be grouped as either *organic* or *inorganic*. Inorganic waste materials, such as aluminum pop cans, come from non-living things. Organic waste materials such as kitchen waste and paper come, directly or indirectly, from living things. Municipal waste consists of both organic and inorganic materials.

Both organic and inorganic waste materials are *degradable*; that is, they will break down given the right conditions. Most inorganic materials, such as metal, break down very slowly, sometimes taking hundreds or even thousands of years. Organic materials break down faster, in days, months or a few years. A succession of living organisms, insects, mould, fungi, yeasts and bac-

What You Will Learn:

- the trends in waste production over the past 20 years;
- the types and quantities of waste that we produce;
- the advantages and disadvantages of simple landfills, sanitary landfills, and incineration;
- the manner in which we in this province deal with our garbage, with special emphasis on what is done in your community;
- the special problems with certain common waste products;
- some solutions to the waste disposal dilemma.

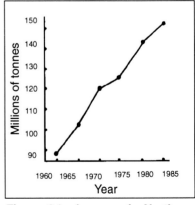

Figure 4.1 - Increase in North American waste production from 1960-1985.

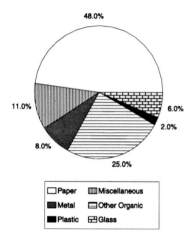

Figure 4.2 - Composition of waste in Newfoundland and Labrador.

teria consume organic waste material and release nutrients to be used again. These organisms are often called *decomposers*, and the process is called *decomposition*. To do their job, most decomposers need oxygen and warm, moist conditions. Any material that can be broken down by living organisms is called *biodegradable*. Given the right conditions, biodegradable material will break down, but large amounts of it can be foul smelling and can attract animal pests.

Some nondegradable products can be made degradable through the addition of certain substances or alteration of some of their components. For example, plastics are derived from crude oil, which is an organic material. However, they are not ordinarily biodegradable because their polymers (molecules) are so tightly bound that they cannot be broken down by decomposers. By adding corn starch to plastic during manufacture, it can become biodegradable (see A Closer Look, page 106). Similarly, some products will break down through exposure to light or they can be altered to do so. These materials are *photodegradable*.

When nondegradable products are made to be degradable, they can help solve the problems associated with the volume of garbage and litter, but they don't necessarily help reduce soil pollution. Their chemical components still break down and are absorbed by the soil.

Analysis:

1. Distinguish between organic and inorganic wastes. In table format, list everyday examples of organic and inorganic wastes, indicating their source and potential impact on the environment.

2. Identify reasons why paper comprises such a large percentage of municipal waste.

Disposing of Municipal Waste

If almost everything breaks down eventually, why is there so much concern about municipal waste? One of the major concerns is that we are producing waste at a rate faster than we can dispose of it. The second concern is that these materials can cause considerable environmental damage during the time between disposal and degradation.

In Newfoundland and Labrador, most municipal waste is disposed of in simple landfills, sanitary landfills or incinerators.

The Simple Landfill

The *simple landfill* is the easiest and most economical method for disposing municipal waste. An area is designated where users simply dump their garbage wherever they can find a spot. It is then buried to the extent possible using a bulldozer or loader. The frequency with which the garbage is buried depends on the size of the population served by the site. For example, the waste at a site serving 5,000 to 20,000 people is to be covered once per week, while one serving less than 1,000 is to be covered once per month. An area that does not receive this treatment is an open dump and is illegal.

There are a number of problems associated with disposing of garbage in this way. Simple landfills are unsightly, and they smell. They can be the source of disease organisms and breeding grounds for pests like rats, which can spread disease. They are fire hazards and the smoke from burning garbage pollutes the air. When garbage is burned, a **toxic** ash may be left behind. In strong winds, the contents of the landfill can blow across the country, littering everything down-wind. As water passes through the dump site, filtering through the garbage and ash, chemicals, some of which are toxic, may dissolve in the water. This process is called *leaching*. If the landfill site has not been carefully chosen, these chemicals can pollute surface and ground waters.

The Sanitary Landfill

The *sanitary landfill* eliminates many of the problems associated with the simple landfill. In a sanitary landfill, the wastes are dumped, spread and compacted in a site where there is a natural depression or where the ground is first trenched to make room for the garbage. Each new layer of garbage is covered with a layer of earth by the end of each work day. It is estimated that about half a

Imagine!

Each of us uses and throws away about 215 kilograms of paper in a year—newspapers, telephone books, magazines, sale flyers, writing and computer paper, and packaging—far too much packaging! Consider that one tree makes about 47 kilograms of paper!

Simple landfill

Imagine!

Disposal of waste except in a designated area is illegal. A new law passed in 1992 increases the maximum fine for improper waste disposal to $500,000.

Toxic: poisonous, can cause illness or death to plants and animals.

101

Monitoring wells: special wells around a dump site in which the water is regularly tested.

Imagine!

Anthropologists working in American landfill sites found that there were annual layers of telephone books representing the time of year when new ones were issued and old ones discarded. Each layer could be used to date the garbage above and below it.

Teepee incinerator

hectare of land is needed to service a population of 10,000 for one year using this method.

The rate of decomposition of waste material in a sanitary landfill site is slow because there is little or no oxygen. Archaeologists working in old landfill sites have found 40-year-old newspapers that are still readable. Complex reactions can produce methane gas, hydrogen sulphide and other gases in a landfill, and toxic chemicals may seep into the ground water.

The ideal landfill site should be constructed well above the water table in an area where there is sufficient soil for covering wastes. It should be lined with a layer of clay or plastic to prevent toxic chemicals from leaching or leaking out. **Monitoring wells** should be built around the site so that ground water can be checked for dangerous chemicals. The methane gas should be vented or collected for use as fuel.

A properly constructed sanitary landfill is a better disposal method than a simple landfill. The litter is better contained and controlled; it is cleaner, less smelly and less susceptible to pests; and it can be used for a longer period than a simple landfill due to the compaction and layering.

When sanitary landfill sites become filled, they are covered. This land is sometimes used for housing developments or recreation facilities such as golf courses and parks. Continued decomposition of the buried garbage results in further settling and in some cases the continued release of toxic substances has caused problems for the new occupants of an old landfill.

Incineration

The burning of wastes is called *incineration*. Proper incineration reduces the volume of the garbage by 80 to 85% by turning it into ash. True incineration is the burning of all waste materials—paper, glass, metals, plastics and organic wastes. The resulting ash and unburned materials are disposed of in a landfill site. Incinerators can be very simple, such as the Teepee or conical ones seen in many communities. More complex incinerators use forced air systems to increase burning temperature and special devices on the exhaust chimney to remove air pollutants.

The smoke from incineration contributes to the overall air pollution of the area. If the temperature of the incinerator is not high enough, toxic chemicals cannot be broken down or may be formed. These toxic chemicals may be concentrated in the ash or released into the air.

Analysis:

3. Municipal wastes can be disposed of in simple landfills, sanitary landfills and incinerators. If you were a town planner in a small Newfoundland community, what method or combination of methods would you recommend and why? What other social and environmental factors not directly linked to the selection might you have to consider?

4. Through research, find out about some of the newer methods employed in the disposal of municipal wastes.

5. All living organisms produce wastes in some form or another. If this is so, why do we have such a problem with municipal waste?

Waste Disposal in Newfoundland and Labrador

Many communities in this province find it very difficult to dispose of their wastes properly. It is a challenge for them to find the funds necessary to develop disposal sites correctly and then to manage them well. Our rocky terrain and shallow soil, which is frozen for so much of the year, hamper the proper operation of sanitary landfills. Therefore, most of our wastes are taken to sites that combine the features of an open dump with those of a sanitary landfill. Most trash is simply dumped, spread and compacted. Soil is used to cover it wherever and whenever possible but rarely is this done adequately.

Because of these problems, and the lack of other options, incineration is generally preferred here. Our human population is spread out, so incinerators can usually be placed so that no community is directly downwind. Although incineration is the preferred method by many provincial municipalities, only about one-third of the disposal sites in this province have incinerators. It costs about $200,000 to build a simple incinerator, and

Selecting a Site for Waste Disposal

In this province, any site for waste disposal, regardless of the method used, should:
- be 300 metres from a road and not visible from it;
- be 1.6 km away from homes;
- have a 60-metre wide area around it of cleared soil to serve as a fire break;
- be fenced;
- be 150 metres away from brooks, rivers or ponds;
- have about 2 metres of mineral soil to help cover the waste;
- have year-round access with a gate;
- have easy access to a body of water for fire control;
- have a portion of the site suitable for car wrecks and scrapped metal storage.

Unfortunately many sites do not meet these criteria.

even then many communities find it difficult to operate them properly.

Hazardous Household Wastes

When we hear the words 'toxic waste' we tend to think of industry as the major producer. However, every day tonnes of toxic wastes from the home find their way to landfill sites or incinerators. There are regulations that prevent industry from disposing of toxic substances in municipal landfills and incinerators. There are regulations that control the disposal of wastes from some institutions. For example, hospitals are required to burn many of the wastes they produce. But there are no regulations for home owners (see Table 4.2). You will learn more about hazardous substances when you study them in Chapter 19.

Imagine!

Every day North Americans throw out enough styrofoam (which contains dangerous CFCs) to fill 10,000 gymnasiums from floor to ceiling.

Substance	Problem	Disposal
Car Battery	Contains lead and is highly acidic	Can be recycled
Antifreeze	Sweet tasting, poisonous	Use up or wash down the drain with lots of water
Motor oil, brake or transmission fluid	Contain poisonous compounds, harmful to wildlife	Take to service station
Lacquer, varnish thinner, and stripper	Poisonous, some are flammable and can cause cancer	Use up or call Department of Environment and Lands
Paint	Contains poisonous compounds	Use up, give away, or remove lid and allow to dry to a solid, then put out in garbage
Drain cleaner	Poisonous, can burn	Use up or wash down drain with lots of water
Oven cleaner	Poisonous, can burn, may cause cancer	Use up and discard empty container with garbage
Furniture polish	Contains poisonous solvents	Use up and discard empty container with garbage
Insecticides, herbicides, and fungicides	Poisonous, can persist for long time	Use up and discard empty container with garbage or call Department of Environment and Lands
Rodent Poison	Poisonous, can kill pets	Call Dept. of Environment and Lands

Table 4.2 - Hazardous household wastes.

Solving the Problem

We cannot continue to throw away our garbage to the extent we are doing. North America has 8% of the world's population, but produces 50% of the world's garbage. In 1989, Canadians produced more municipal waste per person than any other country in the world—and we spent $1.5 billion per year disposing of it.

While many larger cities are running out of landfill space, public opposition to new landfill sites and incinerators is growing. It costs millions of dollars to clean up these old sites when leaching and settling occur.

Reduction of municipal waste can only be accomplished by *integrated waste management*, which is the responsibility of all members of the community.

REDUCE waste by:
- Selecting products that will last.
- Avoiding products with excessive packaging.
- Selecting products with recyclable packaging.
- Packing groceries into as few bags as possible or, better yet, using your own bags.
- Choosing "environmentally friendly" products such as toilet paper and tissues made from recyclable paper.
- Compacting bulky items.

REUSE by:
- Buying drinks in containers that can be returned.
- Fixing things that are broken rather than throwing them away.
- Giving used clothing to others.
- Using ice cream tubs and glass bottles for storage.
- Taking magazines and books to a hospital or senior citizens' home.
- Renting something you only need occasionally.

RECYCLE means to:
- **Compost** your kitchen waste.
- Use a recycling centre if you can.

RECOVER means to:
- Reclaim materials or energy from waste that cannot be reduced, reused or recycled.

Components of integrated waste management include:

Education - Improve knowledge, attitudes and actions

Laws - For beverage container return and recycling

Reuse - Returnable bottles, boxes, etc.

Reduction - Products with less packaging

Recovery - Use of waste for fuel

Recycling - Paper, plastic, metal

Composting - Use of kitchen waste

Product design - Biodegradable plastic

Compacting - Reducing volume of waste

Last Thoughts

Although we have identified 4 Rs (see margin) for waste management, reduction of waste may be the most important by far. We produce too much garbage! Energy recovery from waste and recycling programs are important,

Compost: organic waste such as kitchen scraps mixed with soil in the presence of oxygen. Breaks down into humus that can be used in gardens.

Imagine!

The largest component of municipal waste in Newfoundland and Labrador is paper—newspapers, telephone books, magazines, sale flyers, writing and computer paper, and packaging— far too much packaging! Each of us throws away about 215 kilograms of paper in a year. Consider that one tree makes about 47 kilograms of paper.

but they can give us false confidence that we are dealing with the waste management problem adequately.

Why do we produce so much garbage? Could it be that the root of our problem is the disposable nature of products, and thus overconsumption? Consider the many products we buy that are designed to be thrown away rather than repaired or reused.

As a society, maybe it is time to question our technological advances in relation to wasteful consumer habits. As individuals, it may be time to consider simplifying our lifestyles and making wiser choices in what and how much we consume. Remember that consumers can wield great power in the process of change! Rather than looking to governments, and to science and technology for solutions, perhaps we should look first at ourselves—at the patterns and habits of our lives and the choices we make each day.

A Closer Look: plastics

Plastic is mainly a petroleum oil product. Its production requires large amounts of energy and results in air pollution. The molecules in plastic are so tightly bound together that they are practically indestructible and will therefore last almost indefinitely. Two approaches are used to make plastic bio- or photo-degradable. One is to add a new ingredient, such as cornstarch, to existing plastic that helps it break down into parts that are too small to see. These tiny bits are lost in the soil. In the diagrams below, each gray block represents a chain of plastic molecules. In the diagram on the right, the white blocks represent cornstarch molecules.

The other approach is to make an entirely new kind of plastic out of material such as sugar. Products made from this material are said to degrade into carbon dioxide and water within a year or so. Both methods help reduce the litter problem, but the first approach still has disadvantages. Adding the cornstarch weakens the plastic, so it takes more plastic to make a bag with strength equal to one without cornstarch. This requires more energy and leads to more air pollution in the production process. After the bag degrades, the plastics are out of sight but are still contaminating the soil. Also, this type of plastic cannot be recycled.

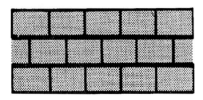

Plastic molecules

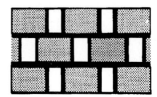

Plastic with cornstarch molecules

106

In 1992, a 'hot' incineration debate was being waged in this province. A United States company, North American Resource Recovery (NARR), was interested in establishing an incinerator in Long Harbour, Placentia Bay. This project was not intended to deal with garbage from Newfoundland, or even from elsewhere in Canada, but instead would transport and burn U.S. waste—3,500 metric tonnes of it a day, 100 million tonnes of garbage a year. This would become the largest incinerator in North America.

The project manager of NARR stated that the project would mean jobs and economic development for Newfoundlanders. He claimed that the company would set up an extensive filtering system to remove toxic pollutants such as ash from smoke, thereby reducing air pollution. He also stated the company's intent to recycle the ash into bricks, concrete blocks, asphalt, even backyard patio stone. The incinerator would also be equipped to convert waste into energy, generating about 100 megawatts of electric power for the area.

However, a group of residents from the community decided to investigate this 'energy-from-waste' project. They formed a group called SNAGG—'Say No to American Garbage Group'—and decided to increase public awareness of their concerns.

SNAGG looked to the U.S. for information gathered from other incineration projects. Discussions with a national environmental group, Clean Water Action (CWA), suggested some serious problems associated with the incineration. Mr. Bob Collins, director of CWA's Waste Management Programs, stated in an interview, "It's pretty terrifying. As an incinerator burns garbage, it releases not only heat, but also the heavy metals—lead, mercury and so forth—that are bound up in the products we use every day. And that either goes up the stack into the atmosphere, or is captured by the... pollution controls and entrapped in the ash."

SNAGG Chairperson Rose Steele stated, "We know that incineration is detrimental to (our) health...This project will produce about 333,000 tonnes of toxic ash each year. What do they plan to do with that? It will release dangerous heavy metals into our air, soil, and water."

In June 1992, the Long Harbour feasibility study was not complete. The situation had not been resolved either in favour of or against the incinerator. Picture yourself as a resident of that community, and think about how you might answer these questions:

- If toxic materials are captured in the ash, what could happen if the ash is used to pave roads, or to make concrete buildings or backyard patios?

- If this project holds such potential for economic benefit, why is a U.S. company willing to give it away to Canada?

- Are the potential risks of this project worth the potential economic benefits?

- Each year this province exports tonnes of waste, such as batteries and solvents from garages, to Ontario or Quebec. What if these provinces decided to close their borders to our waste?

- Many types of land use (pig farms, refineries, penitentiaries) are considered undesirable by those living near them. Should government simply reject these projects if there is strong public opposition to them?

Analysis:

6. *Review your family's wastes disposal practices under the following headings: a. disposal of yard wastes; b. disposal of household toxic wastes; c. disposal of organic wastes; and d. recycling practices. Design a household waste management program that would help reduce the amount of waste that your family produces.*

7. *Do you think that recycling will create jobs or result in job losses? Explain your answer using both local and global examples.*

8. *It is said that recycling can save energy. How is this possible? Use your answer to help outline how recycling the following might result in net energy savings: a. paper, b. aluminum cans, c. plastics.*

9. *What is integrated waste management? Choose one component of integrated waste management and describe a program you might use to introduce it to your community.*

10. *Churchill Falls hydro power is transmitted to the eastern seaboard of the United States. Some of this energy is used to manufacture consumer goods. The wastes from these products may be returned to Newfoundland and burned for energy. What do you think of this energy path?*

Activities: Test Your Trash

1. Bury a biodegradable and regular garbage bag in two buckets of soil. Keep them warm and damp. Measure the rate of breakdown of the biodegradable bag and compare that with the rate of breakdown of the regular bag. Compare these results with the breakdown of a photodegradable plastic bag exposed to sunlight.

2. Analyze cafeteria wastes for one day or one week. Report your findings to the school and develop a fact sheet to be passed out to students suggesting ways of reducing wastes.

3. Redesign a package for an overpackaged product. Send your design suggestions to the manufacturer.

4. Find out about the difficulties associated with recovering and recycling waste resources in your area. Seek out information from a recycling company, your town council, your MHA, or the Department of Environment and Lands.

Students in Action – managing your own garbage

- Practice the 4 Rs— Reduce, Reuse, Recycle and Recover.
- Help your school or community establish a recycling program.
- Write to your MHA recommending laws requiring returnable bottles.
- Develop a proposal and submit it to the school administration for setting up a compost in your schoolyard.

Conducting Your Own Case Study

ENVIRONMENTAL ENCOUNTER

1. Seek information (through research or interviews) about the disposal site near your community and note the following:

   a. How long has the site been in operation?

   b. Why was the site located where it is?

   c. Was there any planning with regard to the layout of the site?

   d. What is the expected life of the site?

   e. Are there any plans ongoing regarding the future of the site and a potential new site when this becomes filled?

   f. Are there any policies in place regarding the dumping of industrial or special wastes?

2. Locate the dump site on a topographic map and make notes about the following:

   a. Determine the drainage basin of the site.

   b. Use your knowledge of contour lines to determine the direction of surface water flow through the site.

   c. Note the route this water takes when it leaves the site.

   d. Use information from weather forecasts to determine the prevailing winds at the site.

3. Travel to your study site and carry out the following activities.

   a. Sketch a map of the site and place the following on your map:
      - areas where specific wastes are dumped (e.g. car wrecks);
      - areas that are no longer active;
      - any streams, lakes or wetlands near or passing through the site;
      - entry and exits to the site;
      - note any loose garbage along the road on the way to the site.

4. If there is a stream moving through the site or near the site, carry out a stream study above and below the site and compare your results.

5. Based on the information collected from your encounter, determine if the site in its present state and location is a threat to the surrounding environment.

6. Suggest ways to reduce the amount of material being deposited at the site you have studied.

7. Present your findings to the local municipal government and/or submit an article to be published in your local newspaper.

CHAPTER 5
Low-Level Flying in Labrador

*The Innu and biologists have identified the negative effects of the flights on various species of animals —
the geese, the beaver, the porcupine, the partridge, all species which the Innu depend on for their
subsistence.*

Resolution on the demilitarization of Nitassinan, 1985

Introduction

We normally associate military aircraft operations with the industrialized areas of the world, not with crystal-clear lakes, mountains, and caribou herds. But since 1957, the people and wildlife of Labrador have had this new influence in their lives.

Low-level flight training (LLT), commonly referred to as low-level flying, involves the operation of fixed-wing jet aircraft at very low altitudes over the hills and valleys of the landscape. Low-level flying is also known as contour flying. Its purpose is to avoid detection from enemy radar by using the terrain as a shield.

There are many social, economic and environmental concerns surrounding low-level flying in Labrador. At the heart of the problem is the question of land ownership and jurisdiction. Land claim negotiations between Canada, Newfoundland and the Innu Nation in Labrador began in 1991, and could take up to a decade or more to resolve. The Innu assert that since they never signed any treaty, they still make the decisions about their land. They say that for governments to proceed with developments during negotiations eliminates their options and is unfair. If developments go ahead, the Innu are worried that by the end of negotiations there will be no natural land left for traditional use. Until the issue of land ownership is resolved, the Canadian and Newfoundland governments believe they have the right to conduct low-level flight training in Labrador.

The Canadian Department of National Defense (DND) studied the possible impact of its low-level flying program on people and the environment. A study of the impact of human activities or developments on the environment is known as an *environmental impact assessment (EIA)*. It is generally required by law for any proposed activity that might have a significant effect on ecosystems and people. The organization wanting to undertake the project is known as the *proponent*. In this case, it is the DND.

One of the most important aspects of an environmental assessment is *public consultation*, a process in which all citizens have a right to participate. The public can comment on the proposed work while the assess-

What You Will Learn:

- the reasons for low-level flying in Labrador, including the link between low-level flying and NATO;
- the conflict between low-level flying and the Innu;
- the meaning and nature of environmental impact assessments;
- the traditional use of the land by the Innu, and their land claims;
- the economic impacts of low-level flying in Labrador;
- the findings of the environmental impact assessment conducted by the Department of National Defense;
- the methods by which the military attempts to minimize the impact of low-level flying;
- the public reactions to the environmental impact statement;
- the special problems associated with noise pollution.

Why Labrador?

The countryside that extends for thousands of square kilometres around Goose Bay is considered ideal for low-level flight training because it is very similar to the kind of terrain that pilots might encounter in parts of Eurasia, an area where a military conflict may likely occur. As well, European countries such as Germany, England and the Netherlands are densely populated areas, and the governments of these nations are reluctant to conduct flight training over their own towns and countryside. Labrador, by contrast, is perceived as empty.

ment is being carried out, as well as when the reports are written. The report describing the assessment is called an *environmental impact statement* (EIS). The proponent, who is responsible for its preparation, must make it available to the public for review and comment.

This chapter will focus on the environmental impact statement prepared by the proponent, the Department of National Defense, for low-level flight training. We will also examine some of the criticisms made of the EIS as part of the public consultation process.

Analysis:

1. *What is low-level flying and why is Labrador suitable for this type of military training?*

2. *What is the main issue surrounding low-level flying in Labrador?*

3. *What are values? What are two conflicting values associated with low-level flying in Labrador?*

Background

The Innu of Labrador

Who are the Innu?

The word 'Innu' is from the Innu-aimun language, and literally translated into English it means 'human being.' The Innu, one of Canada's native peoples, were subdivided by European settlers into two groups–the Montagnais, dwelling mainly in the southern regions of Labrador and Quebec; and the Naskapi, living in the northern interior region. The northern coastal region of Labrador is occupied by the Inuit, a different group of native people. It is important to note, however, that the Innu themselves do not recognize any subdivision of their people or their land. The Innu refer to the land in which they live as 'Nitassinan,' or 'our land'.

For centuries the Innu moved throughout Labrador and into Quebec to find what they needed to live and survive. Many Innu now inhabit the community of Sheshatshiu but the travels of their ancestors followed a strict yearly cycle. In the spring and summer, they fished, hunted small game and gathered what nourishment the land could provide. In the fall and winter, the Innu hunted caribou, known to them as Atiku (pronounced Atik"). During most of the year, the Innu ate beaver, otter, fish, bear, muskrat, porcupine and rabbits.

This lifestyle was a difficult one. It was not uncommon for the Innu to suffer starvation on their long treks because the food they were after, such as caribou, was not where they expected to find it. Yet when they were successful at finding their quarry, they considered it to be a great gift. Animals and other features of the natural world took on spiritual significance.

From the beginning of the 18th century the Innu were gradually drawn into the fur trade with Europeans. The area now known as Sheshatshiu became one of the more important trading posts, and many Innu would trade there with these new people in the summer.

112

Sheshatshiu grew, the wanderings of the Innu slowly diminished, and their lifestyle changed under the influence of the new settlers.

Some Innu had always resisted large-scale industrial developments that interfered with their hunting-gathering way of life. But in recent years, these numbers have swollen as more of the Innu have begun to assert themselves. The Innu are now striving to allow their traditional culture to re-emerge and are persuading the rest of the world to recognize what they value. One of the most recent influences on their lives is the low-level flight training over the remaining areas where they still hunt, fish and trap. Their new voice is clearly heard in this reaction to the low-level flying:

> *We, the Innu people of Nitassinan, from St. Augustin, La Romaine, Natashquan, Mingan, Davis Inlet, and Sheshatshiu, unanimously oppose the use of our territory by the military and we will use any peaceful means at our disposal to put an end to the flights and their abuse of our people and our land.*
> (Resolution on the Demilitarization of Nitassinan, May 31, 1985)

Lengthy stays in the country are part of the Innu lifestyle. The Innu word for country is 'Nutshimit'.

Analysis:

4. *Research the changes that may have taken place in the Innu lifestyle after contact with the Europeans.*

5. *Do you agree or disagree with this statement: "The loss of native cultures is a loss for all Canadians." Support your answer.*

6. *Regardless of the impact study, do you think the Innu's response to low-level flying would change?*

The Military Installation at Goose Bay

As a member of *NATO*, Canada states that it has a responsibility to maintain a modern, well-equipped armed force. NATO is the North Atlantic Treaty Organization, an alliance of fourteen European countries, Canada and the United States, who provide military support to each other. This alliance was made after the Second World War to prevent any one country from starting hostilities. The logic was, that if any country became aggressive towards another, the NATO countries would work together to stop the aggressor.

The military's plans for CFB Goose Bay involved two components. Component 1 would see the stationing of 2,150 military people in Goose Bay and the use of 94 aircraft. Around the year 2000, Component 2 was scheduled to begin. This involved the development of a proposed NATO Tactical Fighter Centre, an off-shore bombing range, as well as many other activities and developments. For this part of the project, 1,280 more military and 420 civilian employees would have been involved, and an estimated $305 million spent on new facilities.

In May 1990, however, NATO cancelled its plan for the tactical fighter centre. This led the Canadian military to cancel Component 2 of its plan.

Although the main benefit of being a member of NATO is related to military actions, there are spin-off benefits as well, such as improved trade relations. There are many other agreements between countries that help trade relations, but they are often enhanced by the close relationships that develop by being a member of NATO. Good trade relations are considered important to Canadian life because they keep Canadian industries active and Canadian workers employed.

The first military airfield was built in Goose Bay during the Second World War through the combined efforts of the United States and Canada. It served as a refuelling stop for aircraft flying to Europe from North American factories. After the war, the base grew in size as it was recognized for its strategic position between the United States and the Soviet Union.

In the mid-1970s, the United States developed bombers with greater flight range. These aircraft could be situated further from the 'enemy', the former Soviet Union (now the Commonwealth of Independent States), and consequently were safer on U.S. soil. Thus Goose Bay lost much of its significance as a military centre.

However, a number of years earlier in 1957, the British Royal Airforce had begun low-level flight training out of Goose Bay. As American activities declined, international low-level flight training slowly escalated in Labrador. It is difficult to practise this type of training in Europe because of the higher population density and the high volume of local commercial air traffic.

By the late 1980s, flight training at Goose Bay was being conducted by five NATO countries—Canada, the United States, the United Kingdom, the Federal Republic of Germany and the Kingdom of the Netherlands.

When it was decided in 1988 to upgrade Goose Bay from a 'station' to a 'base', its new status meant further growth. This change was part of a Canadian government plan to steadily increase the level of flight training in Labrador.

Analysis:

7. Predict four economic benefits that may have resulted from the expansion of the Goose Bay facility into a NATO Tactical Fighter Centre. Identify four economic costs of the proposed development.

The Goose Bay EIA

Once it became apparent that the Department of National Defense wanted to expand the low-level training activities around Goose Bay, there was public opposition and an environmental impact assessment was required.

Any EIA requires that the area in which impacts might occur must first be defined. This is known as the *study area*. Figure 5.1 shows a map of the study area and the areas involving the two components of the planned military activity. Component 2 was later cancelled, during the process of the environmental impact assessment.

Study Area

The study area included large portions of all ecoregions in Labrador except ecoregions 1, 2, and 10. Review the characteristics of these areas from Chapter 2, pages 78 and 79. Generally, these ecoregions have long, cold winters and short, cool summers. The ecosystems are relatively simple, and are therefore quite vulnerable to disturbance. The study area included forest, barren, wetland, and coastal ecosystems.

Because many different species of wildlife were present in the area, the impacts on each of them could not be studied in detail. Therefore, the proponent studied "valued ecosystem components." These species were 'valued' based on the following criteria: their abundance or status within the study area; public concern; professional concern; or economic importance to the people in the area.

It is also important to note that the study area is not restricted to Labrador, but extends into Quebec, affecting Innu people inhabiting areas there as well.

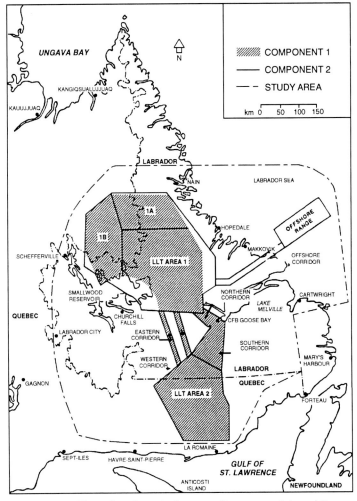

Figure 5.1 - Map showing study area, Component 1 and 2.

Valued Ecosystem Components

The species or groups of organisms that the study focused on included:

- vegetation
- moose
- black bear
- muskox
- red fox
- beaver
- geese
- osprey
- seals
- whales
- freshwater and anad- romous fish
- caribou
- polar bear
- wolf
- marten
- lynx
- gyrfalcon
- bald eagle
- ducks
- seabirds
- marine fish
- peregrine falcon

Wildlife species were selected based on abundance, concerns of the public, professionals and natives, and their eco- nomic importance to the area.

Levels of Impact

- A *major impact* meant that a population may decline or change its distribution and it might stay that way for several generations.
- A *moderate impact* was one in which part of a population might be reduced or influenced to move, but the whole population would remain relatively stable.
- A *minor impact* was one in which a part of a population might be affected for a short period, but the whole population would not be affected.
- A *negligible impact* was one that would be similar to the impact of normal environmental irregularities.

DND divided the study into two parts: one looking at the impact on the natural environment and the other on the human environment. The study ended with a description of the *mitigation*, or the steps the military would take to reduce any negative effects of the low-level flying.

Impact on the Natural Environment

The EIS described four levels of impact on the environment: major, moderate, minor and negligible (see margin).

The military identified a number of elements in their planned programs that might have environmental impacts. They were:

- *Noise* - The study focused on the effects of noise on the immense George River caribou herd, which moves throughout the northern part of the study area. It also considered the smaller herds which stayed within the study area, such as the Red Wine Mountain herd, the Mealy Mountain herd, and the Lac Joseph herd. The EIS stated that the flight training would likely have a greater impact on the smaller herds than it would on the George River herd. Little was known about the long-term effects of this kind of noise, so the aircraft would try to avoid sensitive wildlife areas. Nonetheless, noise generated by low-level flying was identified as the primary source of disturbance for caribou.
- *Aircraft exhaust* - The EIS stated that pollutants from aircraft exhaust would be within air quality standards for the region.
- *Fuel dumping* - During certain evasive or emergency manoeuvres, military aircraft are required to dump some of their fuel. The EIS mentioned that although this might have some severe isolated effects, the overall impact on the environment would be negligible.
- *Dispersal of Chaff* - Chaff refers to tiny strips of metallic material that aircraft release in their wake to confuse the laser guidance systems of missiles. The EIS said that chaff might have negligible impacts on fish and smaller wildlife.

- *Weapons* - The military intended to keep their activities involving weapons away from animal concentrations. The EIS stated that the effects on wildlife would vary from minor to negligible depending on how long it took for the animals to learn to avoid live practice ranges.
- *Aircraft Accidents* - These included impacts with birds and actual crashes. At worst, the impacts of these incidents were predicted to be minor.

The study forecast that the overall impact of the flight training and the activities of the Tactical Fighter Centre on furbearers would be minor even at peak levels of operation. Furbearers include marten, lynx, red fox and the beaver. Impacts on fish were expected to be minor. Consequences for birds of prey like peregrine falcons, gyrfalcons, osprey and bald eagles were predicted to be negligible except in certain areas with heavy aircraft traffic. Existing and potential impacts on waterfowl were thought to be negligible to minor.

Impact on the Human Environment

The Department of National Defense recognized two distinct areas of impact on the human environment. The first would relate to the expansion of the base at Goose Bay; the second to the increased low-level flights over areas used by the Innu.

The EIS stated that the overall impact on the Goose Bay area would be positive because of the boost to the economy generated by the added activity. However, it added that current problems associated with alcohol, drug abuse, violence and diseases might become worse with the rapid influx of people.

DND predicted that the job prospects for many residents of smaller communities would not improve unless they had opportunities for appropriate training. Therefore, these people would continue to rely heavily on the use of renewable resources for a livelihood.

The EIS indicated that competition for the use of renewable resources would increase. This would most hurt the residents of smaller communities like Sheshatshiu because of their reliance on wildlife resources.

Imagine!
The George River herd is the largest in the world, with over half a million caribou.

Sound Levels	
Threshold of Hearing	0 Db
Rustling of leaves	20 dB
Quiet whisper (1 metre)	30 dB
Quiet home	40 dB
Normal conversation	60 dB
Average car (5 metres)	70 dB
Loud singing (1 metre)	75 dB
Average truck	80 dB
Subway (inside)	94 dB
Kitchen gadgetry	100 dB
Power mower	107 dB
Pneumatic riveter	115 dB
Amplified rock and roll music (two metres)	120 dB
Jet plane (30 metres)	130 dB

R. Murray Shaffer,
The Book of Noise
Threshold of Pain 110-130 dB

Table 5.1 - Sound levels of everyday activities.

Imagine!

Employers are required by law to provide ear protection to jack-hammer operators. A jack-hammer creates sound at 92 dB.

Decibel: a unit of measure for sound levels.

Mitigative Measures

DND established a number of new positions at the base in Goose Bay to help deal with the problems associated with environmental and human impact. These positions include an environmental officer, a community liaison officer, a social worker, and a native liaison officer. These staff meet with local people on a regular basis to ensure that the mitigative measures are working.

The study looked at the impacts of noise on people. Excessive noise was recognized as a problem that contributed to hearing loss, general annoyance and stress. The EIS stated that noise levels of less than 70 **decibels** are not harmful to hearing even if a person is exposed to them for long periods. The study claimed that sound levels from most of the flight activity would be around 70 decibels, and in most areas exposure to this level of noise would be for brief periods. However, the report stated that residents near the air field would be exposed to sound levels of up to 100 decibels.

The EIS concluded that the physical impact of the noise on most people would be negligible, but the psychological impact was a real concern, and was declared as such by the Canadian Mental Health Association.

Mitigation and Monitoring

Mitigation or *mitigative measures* refers to the action that the proponent would take to reduce or minimize the harmful impacts of its activities or developments. *Monitoring* refers to the continual review of the activities and developments to see that the predicted impacts and the actual impacts are similar, and also to establish that the mitigative measures are working.

The EIS said that potentially harmful effects would be minimized mainly through avoidance. Because they could not be sure of all the negative impacts on many parts of the natural and human environments, the mili-

A Closer Look: measuring sound

The unit used to measure the intensity of sound is called a decibel (dB). A decibel is one-tenth of a larger unit, the bel, a measurement named after our famous Canadian inventor of the telephone, Alexander Graham Bell.

The normal ear can perceive sound at as low an intensity as zero decibels. Ordinary speech measures about 60 decibels. Have a look at Table 5.1 on page 117 for the noise levels of a number of our day-to-day activities.

The NATO jets in Labrador fly as low as 30 metres, creating a noise of almost 130 dB. The World Health Organization states that 110 to 130 decibels of sound marks the threshold of pain. Can you remember a time when a sound actually hurt your ears? Can you see how noise could be considered a form of pollution?

tary intended to stay away from sensitive wildlife areas and areas being used by people. They planned to set up communications between local people and wildlife officials so all groups could be informed of the locations of activities and wildlife concentrations. The military recognized that this would be difficult.

DND also undertook a separate study to look at ways of reducing the impacts of noise in areas near the airport. The Urban Aerodynamics Noise Impact Study, CFB Goose Bay—more commonly known as the Spruce Park Noise Study—offered several recommendations which have since been implemented.

Analysis:

8. *The caribou is considered a valued species in the Environmental Impact Statement. What do you think are some of the economic and cultural values of the caribou to the Innu?*

9. *Summarize those elements of low-level flight training that might have an impact on the environment. Which one do you think should be of greatest concern? Give reasons for your answer.*

10. *What methods did the military recommend for minimizing the harmful effects of low-level flying? Do you think these methods can be effective?*

11. *Environmental impact assessments are generally required by law for any proposed activity that might have a significant effect on ecosystems and people. Based on what you already know, what might be some of the environmental impacts to be studied before the following activities could be carried out:*

    a) *A hydroelectric development;*

    b) *The development of a new municipal waste disposal site?*

Public Criticisms of the Goose Bay EIS

Two of the people who were significantly involved in the public consultation process were Peter Armitage and Dr. John Kelsall.

Peter Armitage, working on behalf of the Innu, compiled criticisms of the EIS by over 25 technical experts and the Innu. Here we will consider a few sections of his report.

Armitage's criticisms of the EIS were mainly concerned with the Innu and their way of life. Here are some of his concerns:

We all live under the same sky, but we don't all have the same horizon.

Konrad Adenauer

• Existing and potential impacts on the Innu culture were ignored. Given the effects that modernization

119

[The environmental assessment process] is a foreign process. We never accepted Canada's jurisdiction, and are participating [in the process] as one way to see if military flight training can be stopped. However, it is not a culturally appropriate process (e.g. the opinion of an Innu elder or hunter is given little weight, because they do not have a Ph.D.), and the outcome is weighed against us. So we will continue to oppose the flight training by going on the runway and by other non-violent means. Canada's environmental assessment process has been criticised on many grounds. The panel only makes a recommendation, that is often overturned.

Daniel Ashini
Director of Innu Rights
and Environment, 1993

Other Criticisms of the EIS

Some scientists pointed out that the EIS did not look at the impacts on natural communities. By focusing on the impacts of a single species or a group of species, the study overlooked impacts on the interactions among wildlife and with habitat.

has had on the Innu to date, he thought it was unlikely that the influence of the increased economic activity in the Goose Bay area would be positive.

- Both DND and the Innu claimed that the same areas were important to both of them. The military would not be able to reduce its impact on these areas sufficiently to prevent interference with Innu activities.
- Armitage felt that the noise generated by military jet aircraft was underestimated. He also pointed out that the military had understated the psychological damage caused by low-level flying.
- He thought that the military's research into Innu land use was inadequate to determine what areas were important to the Innu.
- Armitage also felt there was a glaring problem with the EIS report's handling of the proponent's mitigative measures, namely avoidance of Innu camps. This was not pre-tested to ensure that it would work properly.

Dr. John P. Kelsall, a prominent Canadian wildlife biologist, dealt with some of the failings in the military study in relation to wildlife. Some of his criticisms were:

- The military did not adequately evaluate impacts on the small, more vulnerable caribou herds of southern Labrador.
- DND claimed to have reviewed all the literature on the caribou of that region. However, Dr. Kelsall failed to find any reference to his own published work on that subject, nor 20 other published works that dealt with the issue.
- He felt that the field work done by the military was insufficient to determine population density and movement patterns of caribou. He said that DND only spent 20 days of flying, at about 500 kilometres per day, to count tracks in the snow. Furthermore, Kelsall argued that the noise of the aircraft may have frightened the animals away and influenced the results.
- There was no reference to how noise affected marten, mink or fox. Fox farmers know that loud noises cause abnormal behaviour in mothers when they are giving

birth. This can lead to the cub's death. Loud noises make these animals nervous and reluctant to mate.

- No research was done on the snowshoe hare which is a large component of many predators' diets. Any decrease in the population of the hare would seriously affect their predators as well.
- Finally, no serious account of the impact of noise, chaff and fuel dumping was done on fish populations and their breeding habits.

Analysis:

12. *The military, as the proponent in this case, conducted many of the studies on the effect of low-level flying in Labrador. If you were a government official who depended on the results of such a study to help you make decisions, would you prefer to use the results of a study done by the proponent or by an independent agency? Give reasons for your answer.*

13. *Based on the public criticisms of the EIS, do you think the study was adequate to help a decision-maker choose whether or not to allow the low-level flying to continue? Why?*

14. *If there has been or is going to be a major development near your area, find out if an environmental assessment was done. If so, obtain a copy of the Enviromental Impact Statement. Use the following guide to review it:*

    a) *What interest groups were consulted during the assessment?*

    b) *What ecological issues were addressed?*

    c) *What social issues were addressed?*

    d) *Were there any areas that you feel were not adequately addressed?*

Activity: Low-Level Flying Public Hearing

Assume that a public hearing is to be held in Goose Bay. The purpose of the hearing is for the military to provide information about its plans for low-level flying and for the public to comment on those plans. From your class, choose people to represent the following positions: a chairperson for the hearing; a military representative; a local business person; an Innu hunter and trapper; a caribou biologist; and an ordinary resident of Goose Bay. The chairperson should get the hearing started, ensure that all participants have a chance to express their points of view, keep things flowing smoothly, and bring the hearing to a close. Each representative should investigate the basis for his/her position as well as possible and express his/her point of view in not more than 10 minutes. Any other student should be free to express his/her view during the hearing. Another group of three to five students should represent a panel of decision-makers who will observe the hearing and then make a decision on the future of low-level flying in Labrador. Argue for your point of view as strongly as you can, but try to respect other people's feelings and points of view.

Last Thoughts

Civilized people depend too much on man-made printed pages. I turn to the Great Spirit's book which is the whole of creation. You can read a big part of that book if you study nature. You know, if you take all your books, lay them out under the sun, and let the snow and rain and insects work on them for a while, there will be nothing left. But the Great Spirit provided you and me with an opportunity for study in nature's university, the forests, the rivers, the mountains, and the animals which include us.

Tatanga Mani (Stoney Indian)

After the initial EIS was published in 1989, the world changed in many ways that caused NATO to rethink the need for low-level flight training. At the same time, the war in the Persian Gulf, where low-level flying was one of the main tactics used, seemed to confirm its military value. The reduced nuclear threat also seemed to emphasize the need for more conventional weapons and tactics. Meanwhile, government budgets became tighter in the early 1990s, and the Innu continued to protest against low-level flying in Labrador. These factors and others led DND to abandon Component 2 of the project, but DND remained committed to Component 1.

The Department of National Defense submitted its EIS to the Federal Environmental Assessment Review Office (FEARO) where it was found to be deficient on 38 points in May 1990. The deficiency statement was revised with the cancellation of Component 2 and was re-issued in December 1991, showing 29 deficiencies remaining.

The reviewers decided that more work had to be done to determine and minimize environmental and human impacts of the project. In 1992, DND was continuing to do this work, and the EIA process was ongoing.

Following the completion of the studies required by FEARO, more public hearings were to be called. After that time, the matter would be referred to the political decision-makers. It would be up to them to determine if low-level flight training would continue in Labrador.

Students in Action – is the issue still 'in the air'?

1. Contact the Innu Nation and DND to get an update on this issue from each of their perspectives. Has anything changed? How likely is it that both groups can be satisfied with the Goose Bay development?
2. If you live near an airport, monitor the noise generated near the end of the runway. Make an audio tape and play it in the classroom. Imagine yourself in the natural environment of Labrador experiencing this level of noise. Write your thoughts and feelings about it in your journal.

CHAPTER 6
Moose and Caribou Management

Beyond the lower slope of the hill seemed to be a solid mass of caribou...
Mina Hubbard (From *A Woman's Way Through Unknown Labrador*)

What You Will Learn:

- the status of the moose and caribou herds throughout the province;
- the economic significance of moose and caribou hunting;
- the major components of moose and caribou management;
- how hunting quotas are set;
- how an aerial census is carried out;
- the role of hunting in moose and caribou management.

Imagine!

All the moose we now have on the island of Newfoundland descended from only six animals that were introduced to the island from New Brunswick!

Moose

Introduction

Whenever we place a heavy demand on a renewable resource, it is very important that we manage it well. If we are to practise sustainable development, we can take from a resource only what is feasible without threatening its well-being or existence. If we harvest trees, or wildlife at a rate faster than the ability of the resource to replenish itself, it will decline and possibly disappear.

Unlike fish and forests, our wildlife resources are not used primarily for commercial purposes. Rather, our management of wildlife populations, such as moose and caribou, is intended mainly to keep them in balance with the habitat that is available to support them. We also allow people to hunt some animals each year (annual licence quota), but people's use of the resource is secondary to keeping the populations healthy.

This chapter looks at how we manage moose and caribou—two magnificent animals that mean a great deal to the people of this province.

Background

Moose

Moose were introduced to the island in two groups: a pair in 1878 at Gander Bay and four more in 1904 at Howley. In 1991, the population of moose on the island of Newfoundland was estimated at around 170,000 animals. This is probably the highest population that has ever existed since the introduction of the species to the province.

Controlled hunting of moose began in 1935 and has continued to the present day. By 1992, over 18,000 moose were being taken by hunters each year.

In Labrador, the moose has a more recent and shorter history. Since the 1940s, moose have moved into some of the river valleys from Quebec, and in 1953 these populations were given a boost by an introduction of new animals from Newfoundland. Relatively small, but growing populations now exist throughout the forested country of Labrador as far north as Nain. The highest densities exist in the Grand River Valley and the Lake Melville watershed. Small-scale moose hunting has been permitted only in southern Labrador where populations are strongest.

Caribou

Biologists estimate that before the Newfoundland railway was built in 1898, there were about 100,000 caribou on the island. The information supporting this is unscientific, however, which makes it difficult to be certain about the number. Nonetheless, market hunting and other factors caused an abrupt decline in the caribou population, which resulted in a complete ban on hunting in 1924. The herds then began to rebuild from a few thousand animals in the 1920s, to about 70,000 by the early 1990s. As the herds grew, hunting was again allowed on a controlled, **non-commercial** basis beginning in 1936, and has continued to today.

Before the 1920s, wolves existed on the island, and caribou were most likely their main food. The wolves and caribou coexisted, with both species maintaining healthy populations. Now, without wolves, the caribou's main predator is people. Bears and lynx also take a few caribou, especially calves, but not in great numbers.

Caribou have always been more numerous in Labrador than on the island of Newfoundland. Labrador caribou have vast areas of good habitat and few people to contend with. However, in contrast to the island, they have wolves preying on their population.

The George River herd—the largest caribou herd in the world—roams throughout the Ungava Peninsula, using habitat in both Labrador and Quebec. The most recent estimate of this herd (1991) is between 400,000 and 500,000 animals. However, when wildlife populations are this large and spread across so much country, it is very difficult (and very expensive) to estimate their size.

Although the George River herd is beginning to wander into southern Labrador, there are other herds that reside there year round. These include the Mealey Mountain herd, the Lac Joseph herd and the Red Wine Mountain herd, to name a few. Compared to the George

Non-commercial: for personal use only, not for sale.

Caribou

Nowadays we see lots of sport hunters. They kill for big antlers. We worry about them, and big developments like hydro dams. We Innu have sustained caribou since long before any Europeans came to our land, without wildlife officers or scientists, and we can still sustain caribou without government interference. We call wildlife officers 'Katshitatikuet' or 'the person who won't let you kill caribou.'

Napess Ashini,
Innu hunter

125

Imagine!

The caribou is the only member of the deer family in which both males and females have antlers.

River herd, these southern herds are small, totalling only a few thousand animals.

Hunting of the southern herds is restricted due to their small size, but 8,000 to 10,000 animals are harvested from the George River herd each year. No hunting quotas are necessary for the George River herd because the hunting demand is small compared to the size of the herd.

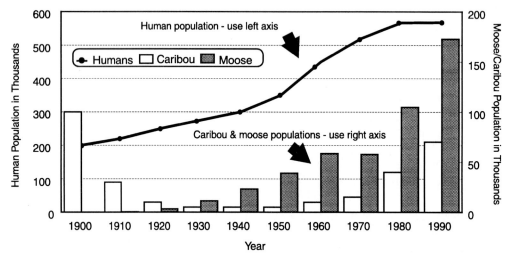

Figure 6.1 - Human, moose and caribou populations by decade since 1900.

Analysis:

1. When were moose first introduced to the island of Newfoundland?

2. Caribou are native to the island of Newfoundland. How do you think the first caribou got to Newfoundland?

3. What might have happened to cause the wolves to become extinct?

4. In the absence of wolves to prey on caribou, what might happen to caribou populations if all hunting stopped?

5. What might be some of the ecological consequences if moose or caribou populations were uncontrolled?

6. Use Figure 6.1 to describe the trends in human, moose and caribou populations since 1900.

7. What happened to the caribou population in the early 1900s? What practice affected caribou populations during this period?

8. Can human, caribou and moose populations coexist without seriously interfering with each other? Give two examples to support your answer.

126

Activity: What's the Dollar Value of Moose and Caribou?

Form working groups, and do a door-to-door or telephone survey in your community to estimate the number of households in which one or more people participated in moose or caribou hunting. Tabulate the number of families that were successful in getting an animal compared to the number that were unsuccessful. Use this data to calculate a dollar value of the meat harvested in your community. Assume the following:

- The average moose provides 160 kg of meat and the average caribou gives 50 kg of meat;
- Moose and caribou meat are worth at least as much as beef in terms of dollars per kilogram.

Assuming a total annual harvest of 2,000 caribou and 20,000 moose for the island, calculate the total economic value of this harvest. (Moose harvests in Labrador are small, and the number of caribou harvested there is not easily calculated.)

Summary of Moose and Caribou Management

There are many ingredients required to manage moose and caribou successfully. For example, basic research has to be done to understand the connections with other species, how much and when populations mix together, how many new animals are born each year, and the condition of their habitat. The number of animals lost each year to predation, crippling loss and poaching must also be determined. Laws have to be created and enforced to protect populations and to allow controlled hunting. And finally, people have to be educated to understand how and why wildlife management occurs and act responsibly when they are using wildlife resources.

Moose and caribou populations in Newfoundland and Labrador are managed based on an *area/quota system*. This means that the province is first divided into sections or management areas, each of which has certain characteristics slightly different from neighbouring areas. Licence quotas are then set for each area depending on whether it is best for the population to increase, decrease or stay the same.

If an area has too many moose, managers will increase quotas to bring the population down in order to prevent damage to the habitat. On the other hand, if the habitat is in good condition, and the area could support

Components of Moose and Caribou Management
- Laws
- Enforcement
- Public education
- Hunter education
- Hunter success
- Quotas
- Poaching
- Legal harvest
- Crippling loss
- Natural mortality
- Age and sex structure
- Habitat assessment
- Population estimates
- Management areas
- Research

Wildlife managers consider the following factors when setting up management areas:
- whether or not each area contains a moose or caribou population that stays mainly within that area;
- the type of habitat;
- the accessibility of the area to hunters;
- the ease with which the boundaries can be identified by hunters.

Imagine!

The day after a caribou calf is born, it can run fast enough to escape from a person on foot. Two or three days after birth it can escape from most of its other predators such as lynx and bear.

more animals, quotas will be kept low to allow the population to grow.

Management of moose and caribou populations in Newfoundland is more intensive than in Labrador because the number of people on the island who use these populations is so much larger.

Although there are strong populations of moose and caribou throughout most of this province, there are not enough animals to satisfy the demand for licences. See Figure 6.2, which shows the size of the moose and caribou licence quotas in comparison to the number of people who look for hunting licences each year (No. of Applications). This chapter will deal with the process used to set licence quotas.

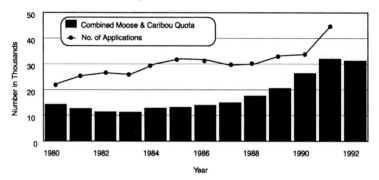

Figure 6.2 - Moose and caribou licence quotas, 1980-1992.

Productivity: number of animals added to the population each year.

Mortality: deaths by natural and other causes.

Predicted hunter success: percentage of hunters that are likely to successfully harvest an animal.

Desired change: the amount by which a population should increase or decrease.

Extrapolate: to predict an unknown figure based on a sample of known data.

Setting Licence Quotas

The managers of moose and caribou populations need the following information to set annual quotas for each management area:
- an estimate of the population;
- the **productivity**;
- **mortality**;
- **predicted hunter success**;
- **desired change**.

Population Estimates

The main method that managers use to determine a population estimate is the *aerial survey*. This involves flying over a management area and counting the animals in it. Most areas are too large to conduct a full census in, so the animals in a portion of the area are counted, and the count for that portion is then used to **extrapolate** an estimate for the entire area. This process is called *sampling* and is very important in many scientific investigations.

Activity: Raisin Census

How might you estimate the number of raisins in a loaf of bread without having to count each raisin? Start by taking a sample—a few slices. What percentage of the loaf do the slices represent? What calculation would you make to estimate the number of raisins in the whole loaf? Compare this activity with estimating wildlife populations.

Counting moose and caribou from the air is much more difficult than determining the number of raisins in a loaf of bread, but the sampling methods used in both cases are similar. Managers start by randomly selecting 4 km^2 blocks from a management area. They wait for the right weather conditions, usually after a fresh fall of snow so that animals show up clearly and their tracks can easily be seen. They generally use a helicopter because of its ability to hover, turn sharply and back up when the observer is trying to make a count of moving animals. A navigator records the information and ensures that the pilot stays within the sample area.

Table 6.1 shows the results of three moose management areas recently surveyed. Table 6.2 shows the year in which a census was previously taken in these areas and the population estimate made at that time.

The numbers and dates used for the remaining part of this investigation have been changed slightly from the source data in order to clearly illustrate three different management strategies: a situation where wildlife managers would want to reduce a population, another where they would want to keep the population stable and one where they would try to increase the population.

> **Imagine!**
> A moose can dive five metres or more to feed on plants growing at the bottom of a pond.

Area	No. of Animals Counted	% of Area Surveyed
20	193 (1991)	12%
24	405 (1990)	12%
34	330 (1990)	12%

Table 6.1 - Recent survey results for three moose management areas.

Area	Population Estimate
20	2117 (1984)
24	3434 (1983)
34	1189 (1981)

Table 6.2 - Previous population estimates.

Analysis:

9. Use the following formula and the data in Table 6.1 to calculate the population in each area:

Population Estimate = No. of Animals Counted
_____ **X 100**
% of Area Surveyed

10. Compare your calculations for the most recent population estimates to those in Table 6.2. What appears to be happening to the populations in each of these areas?

11. What factors might be involved in the changes that have occurred?

Moose as seen in an aerial survey

The cost of doing an aerial survey has become very high. As a result, managers are unable to conduct a census as often as they would like. With almost 80 different management areas (moose and caribou combined), this can mean 10 years or more between censuses for any one area. Managers must therefore rely on other information gathered between aerial counts to determine how populations are doing.

This information is known as *trend data* because it indicates the 'trend' of the population—whether it is growing, declining, or staying the same. Trend data is a measure of *relative abundance*, while an aerial survey is a measure of *absolute abundance* (see pages 70-71). The trend data in this case comes mainly from hunters, and it is essential for successful management.

Each hunter is required to submit to the provincial government his or her **licence return** and the jawbone of the harvested animal. On the licence return, the hunter indicates the number of moose seen and the number of days spent hunting. When this information is gathered from many hunters in a given area, it is used to indicate the trend of the population in that area. If the average number of moose seen per day is decreasing, it is a very strong indication that the moose population is also decreasing. From many years of collecting this data, biologists have found that the number of moose seen per day per hunter is a very reliable tool for evaluating whether a population is increasing, decreasing or remaining the same. Figure 6.3 shows the average number of moose seen per hunter per day on the island of Newfoundland for the past 25 years.

Licence return: the portion of a hunting licence that is completed by the hunter after the hunting season and returned to government.

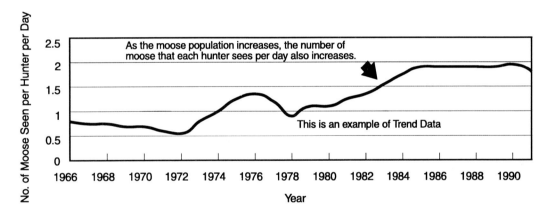

Figure 6.3 - Average number of moose seen per hunter per day.

Analysis:

12. What was happening with the moose population of Newfoundland during 1970, 1980, 1990?

The jawbones provided by hunters go to the wildlife laboratory where a tooth is extracted, ground down and examined under a microscope. Like the rings in a tree, layers in the tooth can be counted to give the precise age of each animal. When the number of animals at each age level is counted for each management area, age pyramids such as those in Figure 6.4 can be constructed. This data is vital to understanding our wildlife population and

Reading Age Pyramids
Although age pyramids are very useful for understanding what might be happening with wildlife populations, biologists cannot rely upon them in isolation. Additional information on rates of increase or decrease are required to interpret the age structure of a species.

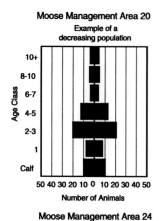

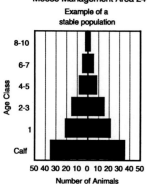

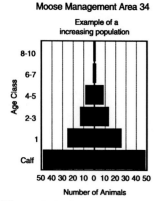

Figure 6.4 - Age pyramids.

relies upon the cooperation of hunters to provide the jawbones.

The age pyramid for Area 20 suggests a decreasing population because there are fewer calves and yearlings than there are 2-3 year olds. There is also a higher than normal number of 8-10 year olds. This is often described as a 'top heavy' age structure. The pyramid for Area 24 shows a stable age structure with an even, step-like, reduction in the size of each age level from the bottom to the top. The pyramid for Area 34 shows a typical increasing population with a much broader base (many young animals) and a greater reduction in the size of each age level from the bottom to the top. It also shows fewer middle-aged and old animals than the stable population.

Productivity

Productivity refers to the number of new animals born each year. Managers determine productivity of moose and caribou mainly through another aerial survey called a *classification survey*. They compare the number of adult cows with calves to the number of adult cows without calves. They can then say, for example, that 15% of the cows had calves, so productivity is 15%. This represents the rate at which the population is growing.

Mortality

While new animals are added to the population, others are dying for a number of reasons. The causes of death are called *mortality factors*. Some of these factors are natural—such as disease, predation, starvation, old age, injury, and severe weather. Biologists must determine, as best they can, the level of *natural mortality* in a population before they can set a quota.

Other mortality factors include poaching, crippling loss, and legal hunting. Unfortunately, poaching and crippling losses are difficult to determine. It is easy to determine the number of poachers who are caught, but it is almost impossible to know how many are not caught and how many animals they kill.

Crippling losses are also difficult to determine because many animals do not immediately die from their wounds and are never found. Biologists have tried asking hunters to indicate on their licence returns the number of moose they find dead. This information has been combined with data collected from field studies that use radio collars. If a radio-collared animal is crippled and then wanders off to die, biologists can find the animal by following the radio signal, and then determine the cause of death. These combined studies have indicated that crippling loss has been between 15% and 30% of the legal harvest in some areas.

Because of the high crippling losses, government has developed an education and testing program to help big game hunters use their firearms properly and kill animals humanely. Hunters must practise their shooting skills, know where to place their shots to cause death quickly, and use rifles that shoot straight. They must also follow the simple rule, *if you are not sure, don't shoot.*

Think About It:
Which would be preferred, taking too many animals from a population, or not taking enough?

Hunter Success

The next piece of information needed to calculate licence quotas is hunter success rates. Hunter success is the percentage of hunters in an area who harvested an animal during the previous hunting season. This success rate is determined from the licence return. Although the hunter success rate varies from area to area, depending on accessibility, density of moose, and so on, it is quite high for most of the province—about 75%.

Desired Change

The last thing to be done before biologists can calculate a licence quota is to decide what they want the population in an area to do—increase, decrease or stay the same. They must ask:

- Are the animals healthy, or are they showing signs of low productivity or starvation?
- What is the condition of their habitat?

Determining Poaching Loss

One study, done in 1985, involved questioning many wildlife professionals throughout the province to get their opinions on how many moose were being poached in their districts. The study estimated that the total loss to poachers was 7% of the population. This was roughly equal to one half of the legal harvest!

If poaching and crippling loss are underestimated, the legal harvest quota may be set too high. This would cause a moose or caribou population to be overharvested. If poaching and crippling loss are overestimated, the quota will be set lower than necessary, fewer hunters will get a licence, and the game population may grow faster than expected.

Imagine!

According to the Canadian Wildlife Federation it is estimated that fewer than one percent of all poaching incidents in Canada are reported.

...there is a belief among enforcement officers that, even though there has been no statistical jump in convictions, poaching is on the rise.

Rick Dophin, 1993

- Are there too many animals in an area or too few?
- What are the impacts of animal populations on people in the area?

With answers to these questions, biologists determine whether they want the population to grow, decline or remain stable. If they want the population to change, they need to decide by how much each year. They also need to determine what the target population (desired density) should be in order for it to be compatible with the habitat. This is because some areas are better able to support large animal populations than others are. The desired change would be represented by a negative number if wildlife managers want to lower a population; it would be a positive number if they want to increase the number of animals.

Analysis:

13. Describe five things that have to be considered before hunting quotas can be set.

14. What is the advantage of using trend data rather than aerial surveys to determine moose populations? How is trend data collected?

Results

From the information provided earlier in Table 6.2 and the calculations based on the data in Table 6.1, it is evident that the population in Area 20 is decreasing; in Area 24 it is stable; and in Area 34 it is increasing. You can now calculate the quotas for these areas by using the information in Table 6.3.

	Area 20	Area 24	Area 34
Population Estimate(P)	1608	3375	2750
Productivity (PR)	0.10	0.45	0.20
Poaching and Crippling Loss (PL)	0.07	0.03	0.04
Natural Mortality (NM)	0.05	0.05	0.02
Desired Change (DC)	-0.10	0.00	+0.10
Hunter Success (HS)	0.67	0.75	0.95

Table 6.3 - Population characteristics of three moose management areas.

Let's look first at what some of these numbers mean. Note that the desired change for Area 20 is a negative number, meaning that the managers want to reduce this population. This is because in this instance, there are too many animals for the habitat, indicated by low productivity compared to the other two areas and an age structure with too many old animals and fewer young ones (see Figure 6.4). In Table 6.3, compare the productivity of Area 20 with that of Area 24, which has an increasing population. The productivity is more than four times higher in Area 24. Nonetheless, the desired change is zero, meaning that the managers want to stabilize the population at its present level in order to prevent it from getting too large. Area 34 has a stable population, but managers want it to increase by 10%. This means that they believe the area can support more animals than it does in this instance.

We know [poaching] is a multi-million dollar business, but exactly how big it is no one can say. It's inevitable that there's going to be a collapse in certain wildlife populations if poaching continues at the current rate.

Colin Maxwell
Canadian Wildlife Federation, 1992

Activity: Calculating the Licence Quotas

Calculate the licence quota for each of the 3 management areas by using the following formula:

Licence Quota = P x [PR - (PL + NM + DC)] ÷ HS

(See Table 6.3 for the meaning of abbreviations.)

Recent Trends

In 1992, both moose and caribou herds on the island of Newfoundland were doing very well. In spite of high demands placed on them by hunters, populations had steadily increased for about ten years. Moose licence quotas had also increased sharply since the mid-1980s in response to growing moose populations, as well as high social and political pressure to reduce the number of vehicle accidents involving moose.

In Labrador, moose populations were slowly increasing in 1992, a trend that was expected to continue. The small southern caribou herds were stable, but the George River herd was declining, although the reasons for this change were unclear.

Coyotes

Coyotes have been spreading throughout the island of Newfoundland since about 1985, although it is too early to say what their impact on caribou or moose might be. In 1992, biologists were predicting that they may have a significant impact on caribou, but there is no evidence of that as yet.

A Closer Look: the role of hunting in population management

Hunting has been a part of human culture for a very long time. We have used the products of wild animals for food, clothing, shelter, tools, bedding, medicine, and as spiritual symbols. We now rely on these products much less than we did, since supermarkets provide most of our needs. Nonetheless, many people continue to hunt for many different reasons.

In a 1987 survey of Canadians, scientists found that 17.1% of all Newfoundlanders and Labradorians hunted in that year. New Brunswick was the only province with a higher rate of participation in hunting. When the time spent hunting was examined, the researchers found that people of this province spent more time hunting than did any other group in Canada.

Many people hunt because they enjoy wild meats, which are very nutritious and low in fat. For some, hunting provides their only source of meat. During times when unemployment and the cost of food are both high, these game meats help stretch a dollar a lot further. Some hunters simply enjoy the opportunity to get outdoors for an extended period each year, often with some of their best friends. The challenge involved with wilderness survival, and the satisfaction gained from success is what motivates other hunters. For many others, hunting is simply a treasured part of the culture that helps tie them to the land or sea and remind them of who they are and where they came from.

While many people practise and support hunting, others oppose it. Some people believe it is unethical to kill wild animals. They say that animals suffer unnecessarily from hunting. Some believe that wildlife is intensively managed just so hunters can have something to hunt. Some people argue that they oppose hunting for sport or for trophies, but don't mind if hunting is done mainly for meat.

The decision of where you stand on the hunting issue depends on your personal values. Individuals will decide for themselves if they like it or not. Yet, regardless of your personal opinion, certain aspects of hunting should be recognized for their importance to wildlife management.

People are so numerous and so widespread that in most natural systems it is practically impossible to allow wild species to interact without our influence. Without controlled hunting, some species would overgraze their habitat. This can lead to malnutrition, starvation, and increased vulnerability to disease within the species. Hunting can also be used to keep the age structure of wildlife populations in a healthy state by helping to remove older animals or surplus young ones.

Most hunting occurs in the fall. By reducing populations before winter, we decrease competition for food and shelter among the animals that remain. This improves the survival prospects of these animals during the most stressful time of year.

The evidence that controlled hunting generally helps wildlife lies in the fact that populations of most hunted species in North America have improved throughout the 20th century. Consider, for example, the history of our own moose and caribou populations as described in this chapter.

Students in Action – hunters be aware!

If you decide to take up hunting, be sure to:

- Practise safe and ethical hunting methods.

- Abide by all hunting regulations.

- Enrol in a hunter-education training course in your community (offered through the Wildlife Division of the Department of Tourism and Culture).

- Learn everything you can about your local wildlife.

- Report hunting violations to your local wildlife officer.

- When spending time in the country, leave it in as good or better shape than when you found it.

Analysis:

15. Describe the ways in which hunting can be used as a wildlife management tool.

16. Without hunting, what methods might wildlife managers use to keep animals in balance with their food supplies? Would it matter if the number of wild animals could not be kept balanced with their habitat?

CHAPTER 7
Wilderness Areas
Bay du Nord

Something will have gone out of us as a people if we ever let the remaining wilderness be destroyed...if we pollute the last clean air and dirty the last clean streams and push our paved roads through the last silence.

Wallace Stegner

Introduction

Protected natural areas are places that have been set aside to allow nature to have priority over human development activities. Such areas include national and provincial parks, wilderness areas, wildlife reserves and ecological reserves. They are areas that people can use, but only in ways that do not alter the environment significantly. The setting aside of natural areas is an important part of a sustainable development strategy.

Sustainable development relies on two basic actions. First, we must develop the earth's resources in a way that considers the rights of future generations to benefit equally from these same riches. Second, we must set aside some areas to be excluded from all development, and within them let natural forces take their course. Setting aside a system of protected areas is an insurance policy, a safeguard, against the mistakes we are bound to make as we develop our resources. We simply don't know enough about how the earth works to avoid these mistakes.

Protected natural areas provide more than a safeguard against mistakes. They protect habitat for many wildlife species, especially those that need some distance from human developments. Parks and wilderness areas sustain vital ecological processes. They help regulate climate, maintain the quality of the atmosphere and water, protect soils and watersheds, provide pollinators for both wild and domestic plants, and serve as nurseries and breeding grounds. Reserved areas also provide settings where long-term studies can help improve scientific understanding. Protected natural areas provide opportunities for outdoor education. They allow for outdoor adventure and recreational activities that are essential to many people's mental and physical well-being. When we establish parks and wilderness areas we are making strong statements about who we are and what we value as a people. Protected areas can be sources of economic growth in themselves by providing jobs and attracting tourists from all over the world. Finally, they are essential for us to maintain the wild beauty of this land.

What You Will Learn

- the definition and significance of protected natural areas;
- the different types of protected areas;
- the contribution protected areas make to overall sustainable development strategy;
- the function of the Wilderness and Ecological Reserves Act;
- the process involved in establishing a protected area;
- the importance of public participation in the establishment process;
- the percentage of land protected in this province and in Canada;
- the factors and decisions involved in designating the Bay du Nord Wilderness Area and the Middle Ridge Wildlife Reserve.

Examples of Protected Natural Areas

Mistaken Point Ecological Reserve on the southeastern tip of the Avalon Peninsula contains some of the oldest multi-celled fossils in the world. **Funk Island Ecological Reserve**, about 60 km east of Fogo Island, has the largest Atlantic Murre colony in the world. **Avalon Wilderness Reserve** includes extensive rolling barrens that are home to a caribou herd of over 5000 animals. **Butterpot Provincial Park** helps represent the Maritime Barrens Ecoregion and includes some forest, barrens, bogs and huge car-sized boulders left by the glaciers.

Protecting Water Systems

Consider the problem of protecting a body of water in the same way that we would protect a piece of land. Rivers and oceans are constantly moving and changing, without clearly definable boundaries. It is interesting to note that despite the difficulties, there are efforts being made to preserve marine and aquatic ecosystems in Canada.

The **Canadian Heritage Rivers System** (CHRS) is a cooperative federal and provincial government program, established in 1984 to give national recognition to important rivers in Canada and to ensure the protection of their natural, historical, and recreational values. The Bay du Nord and the Main Rivers have been nominated to the CHRS and will be managed by the province.

The **Canadian Parks Service** has a **Marine Parks Policy**, and is attempting to establish a national marine park program similar to terrestrial national parks, to protect marine resources and the scenic, aesthetic, and recreational values associated with them.

Around the world, natural landscapes are disappearing as cities and towns grow, as industrial development continues, as wild land is converted to agricultural production, as roads and highways push through, and as wetlands are flooded for hydroelectric developments. Here, in one of Canada's least developed provinces, we may find it difficult to think of losing our natural landscapes—we seem to have so many. We tend to think that much of the island and most of Labrador is virgin wilderness. But, in fact, very little of it is protected. About 4% of our wilderness has some form of protection, but less than half of that is well protected. A great deal of our land is leased to industry. For example, 85% of the forested lands on the island is leased to forest companies. Unfortunately, once it is gone, our wilderness is gone forever. As Alf, the Alien, once said in his television show, "Public lands aren't like pizzas. You can't call up and order more."

People who have seen the human-dominated landscapes of Europe, the Far East and much of the U.S. are overwhelmed by the natural beauty of this province. Already this province is recognized as having some of the world's most important natural areas. The barrens where our caribou herds thrive contrast with many other places where these habitats are endangered. Our seabird colonies—Funk Island near Fogo, Cape St. Mary's, the islands in Witless Bay, Hare Bay and Sandwich Bay—are world-class natural wonders. Our pristine lakes and rivers, and our outstanding coastal scenery attract growing numbers of adventurers and travellers from other countries. The Torngat Mountains of northern Labrador are considered by many to be one of Canada's most spectacular mountain zones, rivalling the famous mountains and fjords of Norway. If we value these places, we must protect them from competing land uses.

Analysis:

1. *Summarize the basic roles of protected areas.*

2. *Select one of the province's national parks and, through research, find out why that particular area was designated a national park.*

Establishing Protected Areas

One means of protecting natural areas is through the Wilderness and Ecological Reserves Act which was passed by the provincial government in 1980.

The stated purpose of the Act is, "to provide for natural areas in the province to be set aside for the benefit, education and enjoyment of present and future generations in the province." It describes two types of areas to be established: wilderness areas and ecological reserves.

Generally, wilderness areas are large and few, while ecological reserves are smaller and in greater numbers. Some ecological reserves are large, however, depending on the natural features that require protection.

The Wilderness and Ecological Reserves Act requires the appointment of a Wilderness and Ecological Reserves Advisory Council. This is a group of experts and citizens whose purpose is to advise government on what areas should be protected and how they should be protected. This group would also hold the negotiations and public hearings that help resolve related land use conflicts as much as possible. The Council works very closely with the Parks Division of the provincial government, which provides technical, financial and human resources to carry out the work of the Council and manages reserves once they are established.

The process of establishing parks, wilderness areas and ecological reserves is slow because these areas have to be selected very carefully (see Figure 7.1). The natural features of the areas being considered must be fully evaluated and compared to ensure that we are choosing the best places to represent the ecosystems of the province. An understanding of ecoregions is particularly important in this process (refer to page 75). Each ecoregion of the province should have a significant portion of it protected in some way. It is important to know what ecoregions are represented in a potential wilderness area and if proposed boundaries for the area include all the interconnected parts that make the area important.

Then the compromises begin. Once protected areas have been established, resources such as trees for pulp

Gannet Colony, Cape St. Mary's Ecological Reserve

**Wilderness areas
are large areas that:**

- people may use to hunt, fish and travel in to experience and appreciate a natural environment;
- allow the undisturbed interactions of living things and their environment;
- help ensure the survival of certain species;
- have pristine or outstanding characteristics.

**Ecological reserves
are areas that:**

- contain unique or representative ecosystems, species or natural phenomena;
- allow for scientific and educational activity;
- protect endangered species and habitats;
- provide standards against which the effects of developments elsewhere can be measured.

141

Figure 7.1 - Flow chart showing the steps involved in creating a wilderness area.

and paper, and minerals in the ground, will no longer be available for commercial use. Therefore, we need to know as much as possible about the potential reserves, including the future uses that might be prohibited, before they are protected. This usually involves negotiation with many interest groups, companies who hold rights to parcels of land, and the general public. This chapter is about the process that was followed to establish the Bay du Nord Wilderness Reserve.

Bay du Nord

Selecting the Area

The first step in selecting a new wilderness area was to determine which large areas remained on the island that did not contain major developments such as roads, hydroelectric projects, mines, or large-scale commercial logging. This led the Council to consider three areas: the centre of the Great Northern Peninsula, the LaPoile area near Burgeo, and the Bay du Nord area, between Clarenville and the Bay d'Espoir highway. The natural features and the level of conflicting land uses in each of these areas were then compared. The Council concluded that the Bay du Nord contained the most diverse and significant set of natural features. At the same time, its conflicting land uses were fewer and more easily resolved.

A study area was then mapped, showing the important natural features of the area. This map was used as the basis for gathering comments from anyone with special interest in the area (see Figure 7.2).

Landscape of the Bay du Nord Area

The study area contained about 4,300 square kilometres of wilderness—a land of rivers and large lakes with wonderful names, many of them Micmac in origin—Medonnegonix, Koskaecodde, Kaegudeck and Meelpaeg. The most important river is the Bay du Nord, over 100 km long and recognized as one of the finest unspoiled river systems in Canada. In 1992 it was nominated to the Canadian Heritage River System (see page 140). About

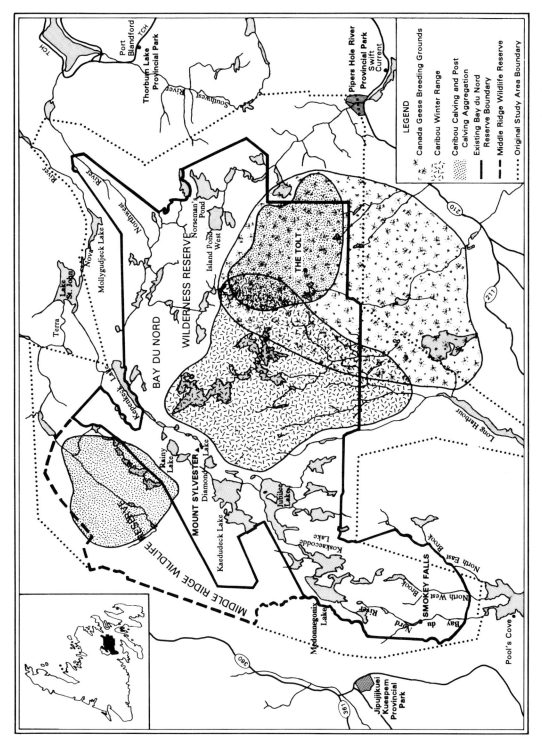

Figure 7.2 - Map of the Bay du Nord area, showing the original study area, the final boundaries of the Bay du Nord Wilderness Reserve, the Middle Ridge Wildlife Reserve and the wildlife habitats within the region.

143

Smokey Falls, Bay du Nord River

Canada goose

15 kilometres upstream from the mouth of the Bay du Nord River is Smokey Falls, a magnificent cascade, above which salmon are unable to migrate. The highest point in the area is Mt. Sylvester which dominates the west, affording splendid views of Rainy, Diamond and Kaegudeck lakes. The highest point in the east is the Tolt, a huge, bald hummock of rock near which the caribou give birth to their calves each year.

Ecoregions Represented

The dominant ecoregion in the reserve is the Eastern Maritime Barrens. It also contains portions of the Western and Central Newfoundland Forest Ecoregions, especially along the river valleys and in the north.

Wildlife

The Middle Ridge caribou herd roams throughout the Bay du Nord area. At 15,000 strong, it is the largest herd on the island. While woodland caribou throughout the rest of Canada are struggling to survive, the Middle Ridge herd is thriving. Moose are abundant in the forested areas and the river valleys. The eastern portion of the barrens area contains extensive wetlands, which provide the best Canada goose breeding habitat on the island. The eastern part of the Bay du Nord region contains some of the best ptarmigan habitat in eastern Newfoundland. The large lakes, with their small islands, provide good nesting habitats for loons and other waterfowl. Five of the rivers in the area contain good populations of salmon, and most of the waters are well stocked with brook trout, ouananiche (land-locked salmon) and smelt.

Analysis:

3. *If you were approached by a tourist and asked to describe some of the places and natural features of Newfoundland and Labrador that make it special, what would you say?*

4. *Distinguish between wilderness areas and ecological reserves in terms of what activities can be carried out in each.*

5. *What are the functions of the Wilderness and Ecological Reserves Advisory Council?*

Analysis:

6. Outline two reasons why the Bay Du Nord area was selected over the centre of the Great Northern Peninsula and the LaPoile area for consideration as a wilderness area.

7. Describe three characteristics of the Bay Du Nord area landscape.

Activity: How much land is protected?

On a clean page of an ordinary exercise book (217 mm x 280 mm), mark out a 34 mm^2 section in one of the top corners. If the entire page represents all the land area of the province, the small square represents all of that area that has been set aside in the form of parks, wilderness areas and other reserves as of 1992 (about 2%). It is only in that small square that we can be assured that activities such as commercial logging, road development, hydroelectric projects and mines will remain absent. Next, mark out a 65 mm^2 section around the smaller one. This represents 7% of the page, which is the amount of land protected in Canada as of 1992. Now mark out another section measuring 85 mm^2 around the last one. This square represents 12% of the land area and is the amount that is suggested as a minimum target for all countries, based on sustainable development policies. Finally, mark out a section 110 mm^2 around the other three. This represents 25% of the land area. Costa Rica, a small country that is much poorer than Canada, has already set aside this much land as parks and reserves.

Write to the Provincial Parks Division and to Parks Canada to find out their plans to establish more protected areas in the province. Ask if we are any closer to the 12% target recommended by the World Commission on the Environment and Development. Explain why you think the amount now protected is too much, not enough, or just right.

Recreation Potential

Most of the rivers and lake systems of Bay du Nord provide a variety of canoeing opportunities for both the novice and the expert. Hiking conditions throughout the barrens area are good, but in the eastern wetlands the going can be difficult. Camp sites are abundant, and the peace and solitude of the surrounding landscape rewards any traveller.

Existing Human Disturbance

It is virtually impossible to find wild areas in the province with no evidence of human use, but within the Bay du Nord this evidence is minimal. The most prominent of such features is the transmission line running from Bay d'Espoir to Sunnyside that marked the southern boundary of the Study Area. About 21 cabins had been constructed within the area, but most of these were illegal.

Criteria for Selecting Areas for Protection

When areas are selected for protection the following factors are considered:

- quality and quantity of natural features;
- rarity of natural features;
- ability of the area to represent its ecoregion;
- size of area;
- level of existing human disturbance;
- present and future conflict over resource use;
- feasibility of protecting natural features;
- proximity to other protected areas;
- similar areas already protected.

Canoeing in Bay du Nord area

The objectives of the Advisory Council for the Bay du Nord reserve were set as follows:

1. To preserve a natural area, with representatives of Newfoundland ecosystems and **biota**, in a pristine condition; in particular:
   - to preserve in a pristine and unpolluted condition the Bay du Nord River and its tributaries, the extensive lakes, ponds and river systems throughout the area, and representative examples of the eastern maritime barrens and forest types;
   - to maintain the Middle Ridge caribou herd at or near the limits of their range;
   - to protect the nesting, rearing, and staging populations of Canada geese;
   - to maintain other species of plants and animals in sufficient numbers to maintain viable long-term populations.
2. To provide continuing opportunities for high quality wilderness recreation, including hunting and fishing, that are compatible with objective 1.
3. To provide a natural area for long-term non-manipulative scientific study leading to understanding of components and ecological processes within the ecosystems present...

Bay du Nord Preliminary Management Plan, 1990

Conflicting Land Use

Much of the northern forested area was within the area leased to Abitibi-Price, the paper company located in Grand Falls. Some of the forested area west of Medonnegonix Lake was also intended for domestic cutting by residents of the Bay d'Espoir area.

Most of the mineral potential of the area was well studied. The only area of significance lay near the west branch of the Bay du Nord River and the upper reaches of the Terra Nova River.

All the larger rivers of the area were considered to have hydroelectric potential, but none were scheduled for development in the near future.

The level of snowmobiling activity, which can be particularly stressful for wintering caribou, varied throughout the area. All-terrain vehicles were used in a few locations, especially along the transmission line in the south and near the mouth of Northwest River where it empties into Northwest Pond.

Establishment Process

After receiving considerable response on the Study Area, the Advisory Council decided that the area deserved further consideration as a potential reserve. They recommended to government that the area be established as a Provisional Wilderness Reserve. This action protected the area until all negotiations and public hearings were concluded. Following these procedures, the Advisory Council would re-evaluate the feasibility of permanently establishing the reserve and present their recommendations to government.

Before hearings were called, regulations were drafted to indicate which activities would be allowed and which ones prohibited within the area. These regulations would be subject to change depending on the comments from participants in the hearings.

It was thought that private land within the proposed reserve would eventually be returned to government, but existing owners could hold ownership until a time when they chose to sell. Government would then buy it for a fair value. Access to private land would continue by aircraft or snowmobile, but not by all-terrain vehicle. Snowmobilers would have to avoid the winter range of the caribou, and aircraft operators would need a permit to land within the reserve. Certain sensitive natural areas were declared strictly off-limits to aircraft landing.

A number of sites near the west branch of the Bay du Nord River and the upper Terra Nova River were thought to have medium to high mineral potential. Yet these areas also contained key natural features for the proposed reserve, namely one of the calving grounds for the Middle Ridge herd and some essential portions of two major river systems. Since mineral exploration is not permitted in a Wilderness Reserve, a compromise was necessary to allow mineral developers to prove the worth of these sites. This led to the creation of the Middle Ridge Wildlife Reserve.

The regulations governing a Wildlife Reserve are more flexible than those for a Wilderness Reserve. They could allow for mineral exploration, provided that it was controlled so as not to hurt the area. After a number of years of exploration, the mineral worth of the area could be better understood. Government and the public could then decide whether the area would be best used as a protected natural area or as a mineral development site. If the decision favoured a mine site, a new arrangement would have to be reached to minimize its impact on the natural features of the area. If the decision was in favour of improved protection, the area could then become part of the Bay du Nord Wilderness Reserve, which would prohibit further mineral exploration. In either case, pub-

Prohibited Activities

The proposed laws and regulations governing the Bay du Nord area stated that the following activities would be prohibited or controlled:
- construction of buildings, roads, etc.;
- logging, agriculture, mining and mineral exploration;
- activities that would alter the quality, course or flow of water;
- activities that would damage wildlife habitat;
- removal of sand, stone or gravel;
- landing of aircraft;
- spray programs;
- use of all-terrain vehicles and snowmobiles.

...protected places are our lifeline to an ecologically stable future. They are places where the forces that animate our planet and make it unique are allowed to operate with minimal interference by man; places where we can wonder and pay respect to other living things and the intricacies of ecosystems.

John Theberge

Conflicting Opinions

Many people did not agree with the continued existence of cabins and aircraft use, describing this as a privilege of the wealthy. Others believed that people's existing rights had to be honoured and that aircraft were not damaging to the environment. A few wanted hunting in the area to stop altogether, while the majority were insistent that it be allowed to continue.

The wilderness that has come to us from the eternity of the past we must have the boldness to project into the eternity of the future.

Howard Zahniser

lic hearings would be required before government made its decision.

For the next stage of the process, government set the locations and dates for public hearings. Public hearings allow citizens and interest groups an opportunity to voice their opinions. If people were unable to attend the hearings, they could submit written comments.

Results of the Hearings and Negotiations

Twenty-one meetings and public hearings were held in the communities around the Bay du Nord reserve, many of which were visited a number of times. People generally supported the creation of the reserve, but were concerned about the locations of the boundaries, some regulations, and some aspects of the way the area would be managed. In a few communities, people worried that they would lose their traditional access to the land.

The Advisory Council had intended to prohibit all snowmobiling in the area, but this met with strong objections. As a result, the area was zoned to allow snowmobile use in specific areas, while protecting sensitive wildlife habitat, especially caribou wintering locations.

In the southeast corner of the proposed boundary, near Piper's Hole River, two portions of the proposed reserve were excluded to allow snowmobile and ATV use in that area. ATV use along the transmission line in the southern region was also important to many residents, so the boundary was moved three kilometres north.

Further west, the residents of Pools Cove and nearby communities wished to cut timber for boat construction and other uses near the mouth of the Bay du Nord River, so the boundary was moved north there as well. A small strip of land on the west side of the area near Medonnegonix Lake was excluded from the area in order to accommodate existing mineral claims.

In the north, the logging company Abitibi-Price wanted continued access to certain areas. In the end, compromises were reached that were reasonably acceptable to both the company and those wanting the area protected. Several portions of land in the northeast were also excluded from the reserve as a result of concerns from Clarenville and Port Blandford area residents.

The Wilderness and Ecological Reserves Advisory Council finally recommended to government in 1988 that it should establish the Bay du Nord Wilderness Reserve and the Middle Ridge Wildlife Reserve. They suggested changes to the boundaries and a management plan reflecting the concerns raised at the hearings. The recommended area covered 3,513 km², including both the Bay du Nord Wilderness Reserve and the Middle Ridge Wildlife Reserve—about 800 km² less than originally proposed.

In 1988, the government made no clear decision on whether to establish the reserves, or to cancel their provisional protection. The Conservative government was defeated in the next election, and the proposal was resubmitted to the incoming Liberal government, who established the Bay du Nord Wilderness Reserve and the adjacent Middle Ridge Wildlife Reserve in 1990.

The reserve is now managed by the Parks Division of the Department of Tourism and Culture as part of a system of protected areas in the province.

Last Thoughts

The people who helped establish this new wilderness area on the island began their work in 1980, but some of them had been working since the mid-1970s towards this goal. It was a goal that required commitment, dedication and a strong sense of fairness to see it through to its conclusion. It required compromise in order to find a reasonable middle ground. The result was the establishment of an area that all Newfoundlanders and Labradorians can be proud of. The Bay du Nord Wilderness Reserve represents the best of wild insular Newfoundland. With this area protected, the total proportion of the province protected as parks or reserves increased to 1.8%, still much lower than most other provinces and the 12% international target. In 1992, the focus of attention for wilderness advocates shifted to Labrador, the site for two proposed National Parks.

We must be willing to forego some economic development in the interests of the preservation of certain of our wilderness areas which are an essential part of the social and cultural heritage...of the vast majority of residents of this province.

...[the Wilderness and Ecological Reserves Act] will enable Government to set aside certain wilderness areas in this province for the benefit and education of our people, thus preserving for all time some of the natural splendours which we have inherited from our forefathers.

Throne Speech, March 1980
Newfoundland and Labrador

How can we expect that precious thin envelope, the biosphere, to sustain life indefinitely ...if we cannot place a tithe or at least 10% of its land area beyond the reach of either inadvertent or deliberate human alteration?

John Theberge

*World experts predict that we have until the year 2000 to choose and protect the areas that are most important. **After that it will be too late**.*

When we find ourselves disturbed by the tragic destruction of rain forests thousands of miles to the south, we should also ponder our own record for resource conservation. While the people of the rain forest region are desperately struggling for basic economic survival, we must remember that our standard of living is light years ahead. It makes sense that our policies on the protection of the environment should reflect that. Before we condemn others for their environmental mistakes, we should be sure that our own house is in order.

As the human dwellers of the earth, we have the power to influence all life on this planet. With this power comes an equal responsibility for stewardship. This was the major concern discussed at the Earth Summit, a 1992 meeting of world leaders in Brazil. By allowing nature some breathing room, we can prove ourselves as responsible members of a larger global community. We can send a clear signal to the world that, in this province, we recognize that we share this planet with many other creatures and that we are willing to balance our needs with those of other species.

Students In Action - making your views known

Public hearings are designed to accomplish exactly what the term suggests—to hear from the public. They are an opportunity for you to present your views, usually to a panel of government representatives or consultants, on a particular issue. The hearings are generally advertised through the media, and offer you an opportunity to either appear in person, or to make a written presentation. Keep an eye out for your chance to make your views known, maybe even as a class project. Here are a few things to keep in mind:

- Even if you are making a verbal presentation, write down your thoughts on paper first to make sure they are clear and ordered. Be sure to leave your written copy with the panel afterwards.

- Be sure to use facts to support your arguments.

- Present your opinions with emphasis and passion. Use your voice to show you believe strongly in the things you are say. (Practice a few times in front of a friend or your mirror.)

- The more times you speak in public, the more comfortable you will feel about it. You will come to realize how important your opinions really are!

Analysis:

8. *Why was it necessary to establish the Bay du Nord area as a Provisional Wilderness Reserve before the decision-making process began?*

9. *What types of activities are prohibited or controlled in the Bay du Nord Wilderness area?*

10. *Do you feel that the process of establishing the Bay du Nord Wilderness Reserve was fair and reasonable to all interested parties? Support your answer with evidence from the text.*

11. *Describe your feelings about the protection of wilderness for future generations.*

12. *Obtain a copy of Canada's Green Plan and summarize Canada's long-term goals for the preservation of natural areas.*

13. *Since 1990, what new protected areas have been developed both at the provincial and national level? Is Canada meeting its long-term goal for the preservation of natural areas?*

A Closer Look: Bay du Nord—then and now

On September 5, 1822, a 26-year-old geologist, William Epps Cormack, with his Micmac guide Joe Sylvester, began a journey of exploration and discovery. Theirs was to be the first recorded trek across the uncharted interior of Newfoundland, from the east coast's Trinity Bay to the west coast's St. George's Bay.

On September 5, 1991—169 years later—two others ventured forth to retrace that historic route. Dan Cayo was an author; Ray Fennelly, a photographer. Their main goal was to record any changes since 1822, and their travels took them right through the heart of the Bay du Nord Wilderness area.

Cormack reflected upon the wilderness of the interior: "The hitherto mysterious interior lay unfolded below us, a boundless scene, emerald surface, a vast basin... It is impossible to describe the grandeur and the richness of the scenery, which will probably remain long undefaced by the hand of man."

Cayo and Fennelly encountered plenty of evidence of "the hand of man." Canals, dams, blasted rock, roads, gravel pits—these were all indications of developments associated with tapping the hydro-electric power of the region's rivers. Litter was abundant, but wildlife was less so...

"Most days we saw at least one or two (caribou), often at close quarters; Cormack's account describes many thousands, sometimes in herds of hundreds. Our sporadic sightings of geese and ducks were a far cry from Cormack's descriptions of bountiful waterfowl. We saw far fewer grouse and partridge, and only occasional signs of bears and foxes. Cormack found loons and beavers on every pond: we found them on every tenth pond at best. We saw no wolves, now extinct in Newfoundland, or pine martens, now threatened on the island."

Don Cayo "Crossing the Rock"
Canadian Geographic, May/June 1992

The Bay du Nord Wilderness area represents a slice of Newfoundland's interior, protected against future developments by people. If we did not preserve this region, what predictions would you make about its landscape and wildlife resources for, say, another 169 years from now?

CHAPTER 8
Newfoundland Pine Marten
A Threatened Species

We know more about the surface of the moon than we do about many of the biological communities we are so rapidly destroying here on Earth.

Peter Raven and Ghillean Prance

Why Preserve Endangered Species?

Plants and animals have evolved and disappeared into extinction since the beginning of life on earth. So why should we be so concerned now for species that become endangered? The reason is twofold. First, the rate of extinction today is 1000 times greater than what has prevailed in the known past; and second, the health of the ecosystems upon which all life depends is threatened when species—the links in the web of life—become extinct.

The variety of plants and animals in an area is known as its biological diversity, or *biodiversity*. The greater the number of different plants and animals in an area, the greater will be its biodiversity. The differences in genetic makeup of individuals within the same species is another kind of biodiversity, sometimes called *genetic diversity*.

We have learned that the stability of ecosystems depends on maintaining all the interdependent species, not just those that seem to provide some obvious benefit to people. The gradual loss of species has been likened to the loss of rivets that hold an aircraft together. Considering the many thousands that are used, it is likely that a few could be lost without disastrous results. But as more and more rivets are lost, the risk of a fatal problem increases dramatically. It would be difficult to point to any one rivet and say, "The loss of that one will cause a crash." The same is true for the loss of species. No one can say for sure what the future impact of the loss of one species of insect or bird might be, but we know that the more we lose, the greater the risk to the planet. Even life forms that we tend not to value highly, such as bacteria, algae and fungi, perform vital ecological roles. We cannot separate the health of our lives from the health of the ecosystems upon which we depend. Therefore, the preservation of biodiversity has become one of the key goals of sustainable development strategies.

In this province, we have already lost the great auk, the Labrador duck, the Newfoundland wolf, and the sea mink. The Eskimo curlew is probably extinct, and the

What You Will Learn

- reasons for preserving endangered species;
- the meaning and importance of biodiversity;
- the meaning of endangered and threatened when we talk about plants and animals;
- the life story of the pine marten and its plight in Newfoundland;
- the special importance of habitat loss to the status of pine marten;
- how scientists determine home range of marten and its significance to determining how much habitat marten need to survive.

Imagine!
While we have given names to only 1.4 million species, we do not know how many there are—10 million or 100 million.

Biodiversity
The tropical rain forest has greater biodiversity than the boreal forest of Newfoundland and Labrador, or other more northern forests, because it has a much greater variety of plants and animals. For example, one tree in the rain forest has been found to hold a number of species of ants equal to the number of ant species in all of the British Isles.

Imagine!
A single teaspoon of soil might contain 10 million bacteria and over 2 km of microscopic fungal filaments.   (*S.B. Hill, 1989.*)

153

Threatened or Endangered?

A threatened species is any native animal or plant species that might become endangered unless people help it survive.

An endangered species is any native species that might become eliminated from all or a significant portion of its range due to human action.

pine marten, the piping plover, the leather-backed turtle, the peregrine falcon and the harlequin duck are all threatened with extinction or endangerment. Through the 1800s and early 1900s the major problem for many species was over-harvesting in the absence of effective game laws. Now, the major problem facing wildlife throughout the world is loss of habitat caused by the invasion of human-related activities, developments and pollution into what were once wild places.

In this chapter we will look at the case history of one of Newfoundland's little known mammals, an animal whose future is in jeopardy—the pine marten. Although it is doing fine in Labrador, the island population is threatened.

Analysis:

1. What are two arguments that can be used to support the preservation of species?

2. What is biodiversity? Why is it important that we preserve biodiversity?

3. There is an animal that has no economic value, and is not pretty to look at, but it is about to become extinct. In a paragraph or two, argue either for or against allowing the extinction of this animal.

The Newfoundland Pine Marten

Past and Present Status

A Geographic Distinction

The Newfoundland pine marten is considered to be a geographically distinct population in Canada, and therefore represents a diversity within a species that is worth maintaining. Genetic testing, which would determine the level of its distinctiveness, was incomplete in 1992.

No one knows how many marten once existed in Newfoundland, but it is assumed they were never very numerous. Although the marten's preferred habitat—old-growth forest—was once more abundant, their food supply has always been rather limited. Before the late 1800s, no snowshoe hares, shrews or squirrels were available to supplement the marten's diet. The meadow vole was, and is, Newfoundland's only native small mammal upon which marten could feed. It is still the marten's main food. When the meadow vole population declines, as it does every few years, the marten's survival is challenged. In spite of that, marten were once far more widespread than they are now (see Figure 8.1).

154

In 1897, a pioneer biologist, O. Bangs, wrote, "...as early as 1870 marten were still common in various parts of the island, but from the increasing... value of the fur [it] is annually becoming scarcer." The ease with which marten could be trapped was already evident by the turn of the century. In addition to this stress, they were also losing their habitat as the old-growth forests disappeared as a result of logging, fire and pest infestations.

Marten were sufficiently scarce by 1934 that the government of the day declared the marten a protected

...the earth is suffering the decline of entire ecosystems—the nurseries of new life forms.
Eugene Linden
"The Death of Birth"
Time magazine, Jan. 2, 1989.

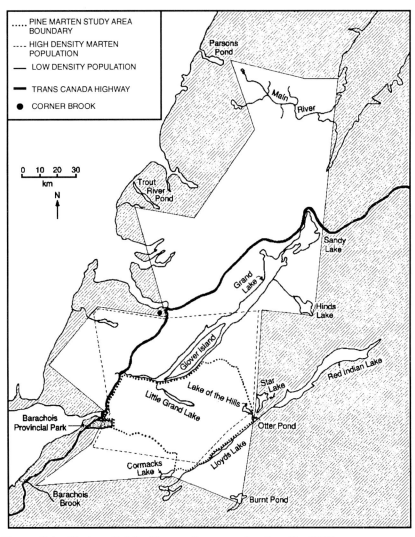

PINE MARTEN STUDY AREA BOUNDARY
HIGH DENSITY MARTEN POPULATION
LOW DENSITY POPULATION
TRANS CANADA HIGHWAY
CORNER BROOK

Figure 8.1 - Marten distribution as it was understood in 1992.

'Variety is the spice of life', goes the saying. Biologists would go further and argue that variety is the very stuff of life. Life needs diversity because of the interdependencies that link flora and fauna, and because variation within species allows them to adapt to environmental challenges. But even as the world's population explodes, other life is ebbing from the planet. Humanity is making a risky wager—that it does not need the great variety of the earth's species to survive....
Halting the assault on biodiversity will not be easy, but there are many actions that governments can take...More money should flow into educational programs that alert people to the irreversible consequences of a loss of genetic diversity.

Eugene Linden
"The Death of Birth"
Time magazine, Jan. 2, 1989.

species. All trapping and killing of marten became illegal. This reduced the pressure on marten somewhat, but trapping and snaring of other species continued, resulting in marten being accidentally killed in traps or snares set for other animals. Before the 1960s, marten still existed throughout central Newfoundland, but the marten's habitat continued to disappear.

In 1973, the provincial government established a 'no trapping or snaring' zone around the Grand Lake and Little Grand Lake area. This provided significant relief to the marten in one area of the province, but outside this area many marten continued to die in traps and snares. From 1970 to 1991, more than 100 marten have been accidentally killed in this way, mostly outside of the area that was closed to trapping. Although averaging only five deaths per year in this way, this level of mortality was very significant for such a small, struggling population.

Disease has also reduced marten populations. In 1986, at least ten marten were lost as a result of an encephalitis infection. The actual number of deaths was possibly higher, since the likelihood of finding dead animals was very low. Productivity has been low in recent years, probably due to the scarcity of prey.

When all limiting factors are considered, the most important one in recent years is habitat loss, due primarily to logging. In spite of this fact, by 1992 no pine marten habitat had been protected.

This chapter concentrates on some recent studies on Newfoundland pine marten and on the work that is underway, or planned, to help these threatened animals recover.

Analysis:

4. Distinguish between extinct, endangered and threatened species. Give one example of each in this province.

5. Summarize the biology of the pine marten under the following headings:
   a. habitat;   b. diet;   c. reproductive behaviour.

6. What are three factors that have had a negative effect on the pine marten population in Newfoundland?

Investigation

A key question that ecologists ask about any threatened species is: what is the minimum required size of the population to ensure its long-term survival? This is important because once a population goes below a certain point, there are too few animals left to interbreed and keep a healthy genetic stock. Another key question is: how much habitat is needed to support this population?

In order to answer these questions, it first had to be determined how much area each marten needs to survive. This area is known as its *home range*. Once the scientists knew how much area each marten requires to satisfy its needs for food, water, shelter and space, they could extrapolate to determine how large an area is required to support a healthy population of marten.

Determining Home Range Size

The basic method for determining home range size in Newfoundland involved trapping animals and placing radio transmitters on them. Their movements were then monitored with the use of sensitive receivers that could pick up the signal from the collared marten.

After the marten was radio collared and released, the researchers moved through the area, usually in a vehicle, to take 'fixes' on its location. Since the radio signal emitted from each marten's collar was unique, the receiver could measure its precise frequency and distinguish it from others in the area. A 'fix' was taken by rotating the receiver's antenna to determine the direction from which the signal was the strongest. Then a compass bearing was taken to match the signal direction and that bearing was plotted on a map. The researchers would then move quickly to a new site and take another bearing on the same signal. After plotting the two bearings, the point where the bearings intersected indicated the approximate position of the marten.

By using this process over a period of several weeks, the researchers were able to plot the area over which each marten travelled. By joining all the outermost locations, they could calculate the size of the home range for each collared animal. Figure 8.2 shows the ranges of two

Pine marten habitat

Radiotelemetry equipment at work in the field

marten as they were plotted over a period of 42 days for one and 125 days for the other.

Ecologists have learned that the average home range for female marten is about 7 km^2, while that for males is 9 km^2. These ranges are about three times larger than those of marten studied elsewhere in North America, probably because Newfoundland marten have to travel further to find enough food.

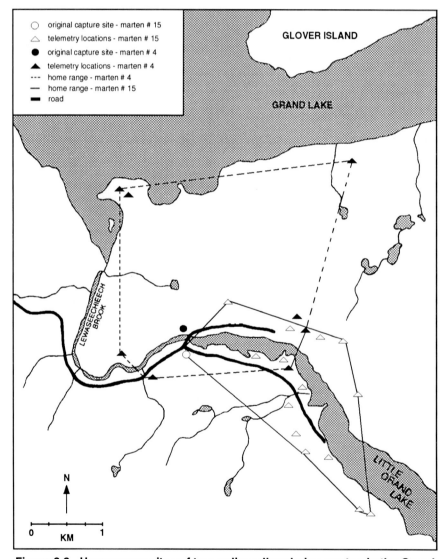

Figure 8.2 - Home range sites of two radio-collared pine marten in the Grand Lake area.

Population Size

Ecologists combined their new knowledge of home range size with other aspects of marten biology and an analysis of how much good habitat existed to support marten. They concluded that there were about 300 marten on the island in 1992. Of these 300, about 180 exist within the high density area shown in Figure 8.1.

But how many marten are needed to ensure their long-term survival? To answer this question, ecologists considered the marten's biology further, including how many females are likely to breed each year, the ratio of females to males in a typical population, and the likelihood of occasional declines in the size of the population. With this and other information, they used a mathematical formula to conclude that the minimum population to ensure the survival of marten in Newfoundland is 237 animals. This population size is considered 'safe' because it would have only about a 5% chance of becoming extinct over a 100-year period.

It is a relatively simple task now to determine how much old-growth forest is necessary to support a population of about 240 animals. Assuming that half of these are males, we know that males each need an average of 9 km^2. We also know that male and female territories overlap to quite an extent, so we can assume that the males and females will generally occupy the same area. If we multiply 120 animals by 9 km^2 for each animal, we know that we need 1,080 km^2 of old-growth forest to support that minimum population.

Although parks and ecological reserves provide protection for some areas of special importance to wildlife, only a small portion of all species can be preserved through these special measures. The future of Canada's native flora and fauna depends more on the will of Canadians to insist that decisions concerning land use and resource development...must reflect sound ecological values.

The State of Canada's
Environment, 1991.

Analysis:

7. Briefly describe how the home range of the pine marten is determined.

8. The male pine marten needs a home range of approximately 9 km^2. Why do you think a pine marten needs such a large home range?

What's Being Done?

Marten have been captured and released into new sites around the province where there is mature forest but no evidence of marten. These sites include Terra Nova National Park, Sivier Island in Notre Dame Bay, the LaPoile River Valley near Burgeo and Main River on the North-

The Recovery Plan

One of the major problems identified by the recovery team was the low level of public awareness and interest in the marten. Because so few people have ever seen a marten, it is difficult for them to want to help it survive. The Recovery Plan therefore suggests major improvements in public education about the plight of the marten. The plan also recommended the establishment of a captive breeding program to provide marten for introduction to new areas and restocking of areas that become depleted.

ern Peninsula. The Terra Nova and Main River introductions appear to have been successful in establishing new populations.

In 1990, a Pine Marten Recovery Team was established to devise and implement a plan to help the marten. The goal of the recovery team is simple—to elevate pine marten populations until they are no longer threatened. Over the next 50 to 70 years, this team would like to help marten populations grow to about 1,000 animals.

Among other things, the recovery team's plan calls for the establishment of three reserves around the island that include suitable marten habitat. The reserves that do not now contain marten would have animals introduced from elsewhere on the island. The team also called for more research on the size and distribution of populations and on experimental logging methods to determine what techniques might be compatible with marten.

The Future

Both of the province's major forest companies, Corner Brook Pulp and Paper and Abitibi-Price, have plans to log within the marten's distribution area in western Newfoundland. Unfortunately, forest management objectives are generally aimed toward production of commercial timber in the shortest possible time, not the maintenance of old-growth forest. However, no cutting has been permitted in the Little Grand Lake area since 1988. Corner Brook Pulp and Paper was also required to undertake an Environmental Impact Assessment on its cutting plans in the Little Grand Lake area. By 1992, their Environmental Impact Statement had not been approved, and the conflict between continued logging and marten habitat preservation was unresolved. Many believe that some logging methods may be compatible with marten preservation, but none has been demonstrated thus far.

Considering the current marten population estimate of about 300 animals and the minimum population needed to ensure long-term survival, we might conclude that the future of the marten population is secure. How-

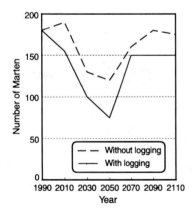

Figure 8.3 - Predicted marten numbers from 1990 to 2110 in the high density area.

160

ever, this would be incorrect. The future of the habitat must be considered as well, and this habitat was still unprotected in 1992. When scientists predict the changes that are likely to occur in the remaining old-growth forest—both natural and human-caused—they foresee a serious problem developing around the years 2020-2060. By that time, they believe there will be sufficient habitat to support only 100 marten, significantly less than the minimum 'safe' population size (see Figure 8.3). The data used to develop Figure 8.3 makes many assumptions about the future and should not be interpreted as fact. Nonetheless, it is part of the nature of environmental science that scientists must predict consequences based on the best information available.

...most people hardly see the point of worrying about insects or plants. But extinction is the one environmental calamity that is irreversible. As these lowly creatures disappear unnoticed, they take with them hard-won lessons of survival encoded in their genes over millions of years.
                                    Eugene Linden

Analysis:

9. A recovery team has been established to devise and implement a plan to help the pine marten in Newfoundland. Describe four components of their plan.

10. Name two natural factors that may affect the remaining old-growth forest in Newfoundland.

11. Obtain the most recent list of endangered species in Canada. Select one species and, through research, find out what is causing it to become endangered and what measures are being taken to protect it.

Last Thoughts

In the world of business, the capital stock of buildings and equipment is the basis upon which an enterprise is built. A steady loss in this capital stock will bring an end to the business. Similarly, plants and animals are the capital stock of the earth, and investment in their preservation is a key to sustainable development. This is especially true since we do not know which spark of life might be the key to curing a serious disease, or to providing a new source of medicine or food.

The preservation of biological diversity is simply a part of the greater goal of protecting the health of our environment. If we allow the marten and other species to disappear, we are clearly not committed to that goal. Such action would instead represent mismanagement of our resources and a setback for sustainable development.

Sharing the Earth
Besides considering the benefits that we might gain from the preservation of biodiversity, do we not have a moral obligation as well—to learn how to share the earth with all other creatures so that our actions do not threaten any one of them?

A Closer Look: what happened to the Great Auk?

On June 3, 1844, on a tiny island near Iceland, two men encountered a pair of nesting birds. The birds died at the hands of these men as they struggled to protect their single egg. Then, one of the men stepped on the egg. And the Great Auk was extinct.

The story of the Great Auk is a painful reminder of our potential for greed and destruction. Once numbering in the hundreds of thousands, the Great Auk was a large flightless bird that nested in colonies during the summer months. The largest known colony in the world was on Funk Island, off Newfoundland's northeast coast.

Early visitors to the shores of North America encountered this small island and its feathered inhabitants. They would launch their boats and go ashore with one thing in mind—to kill Great Auks for food. Some were eaten fresh, while others were salted in great quantities to ensure there was meat on board ship for the return voyage to Europe. The birds' eggs were taken as well.

This slaughter continued until the early 1830s, after which not one Great Auk could be found on Funk Island. Not another has been found anywhere since that desperate struggle in 1844 to protect a single egg.

Students In Action – making space for wildlife

There are a few simple things you can do, either on your own or as a class, to improve the environment in your community and to assist wildlife at the same time.

- Clean up litter such as fishing nets and line, plastic bags, beer cans, broken glass, hooks, and other things that might be harmful to wildlife (and people!).

- Don't ride your ATVs over sensitive areas such as bog—you'll be destroying important wildlife habitat.

- Grow a garden with shrubs, flowers and plants to attract birds and other wildlife. Plant a row of shrubs or trees along a fence.

- Build birdhouses, birdbaths, birdfeeders. Be sure to keep your feeders stocked with food in fall and winter.

- To keep wildlife from making their home in or too near your home, make sure you keep your garbage well-covered.

- Come up with your own ideas for making space for wildlife!

CHAPTER 9

Forest Management in the 1990s

When sun rays crown thy pine clad hills...
(From "The Ode to Newfoundland," by Sir Cavendish Boyle)

What You Will Learn

- the importance of forests to our province;
- past and present harvesting methods;
- the relationship between timber management, integrated forest management and silviculture;
- a variety of silvicultural techniques;
- insect pest infestations and methods of control through integrated pest management;
- some of the problems with large-scale use of insecticides;
- the story of the spruce budworm, which infested Newfoundland forests in the 1970s and 1980s.

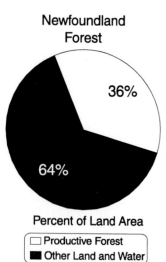

Newfoundland Forest

36%

64%

Percent of Land Area

☐ Productive Forest
■ Other Land and Water

Adapted from: 20 Year Forest Development Plan, 1992

Figure 9.1 - Productive forests in Newfoundland.

Introduction

To the rest of the world, Canada is known as the 'forest nation'. Many countries see us as a huge, tree-covered wilderness that is the source of many of their wood products. These products accounted for 17% of all Canadian exports in 1988.

Within Canada, our relationship with forests has many facets. In 1990, Canada's forest products were valued at $22.2 billion, and the forest industry was providing 670,000 jobs. But the value of forests to most Canadians goes far beyond this significant economic return.

In 1991, the forest and related industries accounted for one in 25 jobs in this province. The economies of some communities, such as Grand Falls and Corner Brook, rely heavily on the incomes of workers in the forest industry.

Although both Newfoundland and Labrador have large areas of forested lands, this chapter will focus mainly on forest management in Newfoundland. The forests of Labrador have great potential, but forest management on the island is much more intensive, mainly because the major paper mills are located there.

The early settlers to this province often used the word 'pine' when referring to any species of conifer. Hence, "thy pine clad hills" suggests more extensive pine forests than we had in reality. About 10% of our forested land may have been pine, but that has been reduced to about 1%, mainly by fires and logging. Nonetheless, Newfoundland still has almost 2.8 million hectares of forests that are considered to be economically valuable—lands that are usually referred to as *productive forest* (see Figure 9.1). Productive forests, which are mainly concentrated in the central and western regions of the island, contain either softwoods (such as balsam fir and black spruce), hardwoods (such as birch), or mixed woods, containing both softwoods and hardwoods. Black spruce and balsam fir forests are the backbone of our forest industry, making up about 75% of all productive stands.

This chapter will describe some of the environmental concerns and management practices surrounding the use of one of our most beautiful and valuable natural resources—our forest ecosystem.

Background

Logging practices have changed greatly since early settlement. The first settlers were likely overwhelmed by the sheer abundance of the New World's forests—a seemingly endless supply of fuel wood and construction material for homes and ships. The abundance contrasted sharply with the wood supplies of Europe, which were then beginning to run short. Yet, the dangerous idea of a limitless supply of wood is only just beginning to change in the minds of many people.

We depend on wood today almost as much as the early settlers did, but for different reasons. Nearly 10% of Newfoundland's total timber harvest still goes into wood stoves and furnaces. But with the rise of iron, steel and a host of other construction materials, the emphasis on wood as building material has shifted to meet other demands. The information age has created an incredible demand for paper, which is higher today than ever before.

Logging Practices, Then and Now

Newfoundland's first loggers brought down the largest white pine trees with axes and crosscut saws. The walls and roofs of many old Newfoundland homes are constructed from great, wide planks sawn from these trees. The logs were taken from the forest by horse-drawn bobsleds travelling over crude snow-covered roads. The diesel tractor first appeared around 1920, but it did not challenge the horse, or other labour-intensive methods, until World War II when there was a severe shortage of people to work in the logging industry.

By the early 1950s, the chainsaw had arrived, marking the beginning of mechanized logging practices. Shortly thereafter, trucks replaced horses forever. What followed these changes was a period of destructive logging practices which did not improve until the 1960s. Bad cutting methods and inappropriate machinery, operating on poorly constructed roads, damaged extensive areas of land. The practice of harvesting only the biggest trees was common, and so was the habit of leaving behind

We Depend on Forests—

- for our economy
- for their life-supporting ecological processes
- for the habitat they provide for wildlife
- for water and soil conservation
- for their influence on climate
- for construction material or fuel
- for scientific study and education
- and for recreation, inspiration and spiritual renewal

Imagine!

Most of the barrens of eastern Newfoundland were once forests. They were destroyed by repeated fires caused by the old coal-fired train engines.

Imagine!

Horses used for hauling wood in the winter were once fitted with snowshoes.

Destructive Logging Practices

- leaving large stumps and tops
- removing the best timber while leaving isolated clumps that would blow down in high winds
- destroying fish habitat through log-driving on rivers and using heavy equipment in streams
- constructing poor roads
- lack of planning
- cutting too close to the edge of bodies of water

Skidder

large portions of these trees as wastage. Rivers were used to float logs from the cutting site to the mill, until this practice became uneconomic because of high labour costs.

Present-day forest harvesting methods have improved immensely since the 1960s. The rubber-tired skidder, although far from harmless, is much less damaging to the forest than the more destructive metal-tracked bulldozer. Logging roads are now designed and built to cause less harm to fish habitat. But the biggest change has been in forest management, which has grown from a fledgling profession into a sophisticated one supported by many branches of science. It is now entering the age of sustainable development. Forest companies and governments alike know that we must continue to improve forest management practices if future generations are to benefit from the rich forest resources of our province.

Analysis:

1. *How many hectares of economically valuable forests exist in Newfoundland?*

2. *Outline five ways in which forests are important to us and the environment.*

3. *Contact your regional forestry office to find out which forest management area your community falls within. How many hectares of economically valuable forests exist in your area?*

4. *Research the role of the forests in one of the following: water cycle, erosion prevention, watershed maintenance, recreation.*

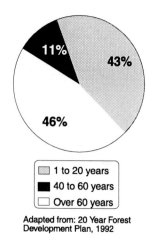

☐ 1 to 20 years
■ 40 to 60 years
☐ Over 60 years

Adapted from: 20 Year Forest Development Plan, 1992

Figure 9.2 - Age of 1992 Productive Forest in Newfoundland.

Forest Management
Meeting New Demands

The process of planning the use of our forests to meet present and future needs has improved greatly over the past few decades. In 1992, a new *20 Year Forest Development Plan* was presented by our provincial government. It described the forest resource base, evaluated existing and future demands on it, and provided a management strategy for dealing with these demands while maintaining healthy forests for future generations.

The plan highlighted the problem of a looming shortage of intermediate-age trees (from 40-60 years old). It states, "If the old forest is harvested too fast and the new forest is not ready, a period of time will occur when insufficient accessible growing stock exists to support the planned harvest." This means a potential wood shortage in the future (see Fig. 9.2).

Forest managers have to help prevent shortages like this, but this is not their only challenge. They must also cope with new sets of demands and concerns. One of these changes is a shift from simple *timber management* towards *integrated forest management* (IFM). Timber management refers to managing our forests primarily for wood products. Integrated forest management means managing forests with many more factors in mind—such as maintaining forest ecosystems and wildlife habitat, maintaining watersheds and enhancing recreation opportunities.

IFM is a challenging idea that requires all groups with an interest in forested land to share their points of view and agree on how it should be managed. Only then can reasonable compromises be reached and a long-term plan be developed. This may sound like a simple process, but in reality it is very difficult.

This type of planning is relatively new in this province, so there are no examples of where it is working. However, the federal government announced in 1992 that an experimental *model forest* program was being established on the west coast of the island. It will be one of ten models in Canada that will try to apply the principles of IFM.

Once an integrated forest management plan is developed, the main forest management tool, **silviculture**, is used to meet the objectives of the plan. Many silvicultural practices involve some kind of tree cutting. Among these are clearcutting, salvage cutting, clearcutting in strips and blocks, selection cutting, and thinning.

Clearcutting: This is the most common cutting method in this province. It means harvesting all the trees in a large area, leaving extensive, open cutovers. Clearcutting is widely used because it requires much less road construction and is therefore the most efficient and economical technique. Salvage cutting is a type of clearcutting in which large numbers of dead or dying trees are cut after

Silviculture: the branch of forestry that deals with the development, cultivation and reproduction of trees.

Cut tree   Uncut tree

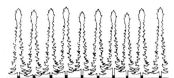

Clearcutting

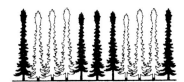

Strip cutting

Selection cutting

Precommercial thinning

167

Imagine!
More precommercial thinning is practiced in this province than in any other province in Canada.

Weed species: plants that compete with or harm the type of trees that are desirable in a forest stand.

Common Silvicultural Practices
- many different cutting practices
- seeding and tree planting
- soil improvements
- thinning and pruning
- salvaging dead trees
- changing species composition
- insect pest and weed control
- controlled burning

...the time has come for forestry in this province to make the difficult transition from controlled exploitation of the natural forest to sustainable development of a managed resource...

20 Year Forestry
Development Plan

Imagine!
Balsam fir forests tend to regenerate naturally without planting, so most replanting in this province is with black spruce.

---

an insect infestation. A salvage cut helps reduce the economic loss and prepares a site for new growth.

Strip and block cutting: Narrow channels or blocks are cut within a forest. The trees left standing provide a source of seed and shelter for new growth. In this province, strip cutting has been practised only on an experimental basis. Block cutting has not been used to a great extent, but small-scale domestic cutting, which occurs in scattered areas, has a very similar effect.

Selection cutting: During any one year, 5% to 10% of a forest is cut. The trees left standing serve as a seed source and shelter for regeneration. Over a period of 10 to 20 years the whole area would eventually be cut. This method is not practised in this province. It is more common in hardwood forests such as in southern Canada.

Thinning: A stand is thinned to reduce competition among the remaining trees. Precommercial thinning (PCT) involves removing trees from a 10 to 15-year-old forest so the best trees are separated by about two metres. This practice allows a forest to reach marketable size about 20 to 30 years faster than a stand without PCT. This technique is widely practised throughout the commercially important forests of the province.

Besides cutting methods, there are many other silvicultural practices. Among the most important are replanting, weeding and pest management. Replanting involves the planting of seedlings to help an area regenerate with the most desirable species. To control **weed species**, foresters in this province apply herbicides that kill unwanted trees and shrubs in immature stands. Foresters also use a wide variety of methods to manage insects, one of which is the use of pesticides.

Some silvicultural practices, such as clearcutting and the use of chemicals to control weeds and pests, are quite controversial. The sight of a large, recent cutover, marred by the tracks of heavy equipment, sparks many different responses in people. While one person might see it as devastation, another would see it as the beginning of a new cycle in the forest. While large-scale clearcutting causes a major ecological change in an area, favouring one new group of plants and wildlife and eliminating others, this type of change can also occur in nature, with or without the influence of people. The effect of clearcutting is little different from a major forest fire, which would also remove old trees and begin a new cycle of growth. In fact, most of our black spruce forests, which are so important to the pulp and paper industry, are a direct result of fires (see A Closer Look). As bad as a recent cutover can look, it will not stay that way for many years. As long as the soil is not eroded by poor road construction, the forest will be restored.

The concern with herbicides and pesticides is similar to the concern with use of other chemicals in the environment. Because of major ecological damage caused by some chemicals used in the past (DDT and others), many people continue to worry about the long-term influence of different chemicals used today. Even when chemical manufacturing companies and government assure us that these chemicals are relatively harmless, many people remain sceptical.

Imagine!
In 1992, 65% of the productive forested lands of the island were leased to two pulp and paper companies.

A Closer Look: forest fires...good or bad?

We tend to think of fires only as destructive and dangerous occurrences, and indeed they can be. But they can also be a natural part of the life of a forest. Many forests depend on fires to some extent for regeneration—to clear out the old growth and start anew. The forests of central Newfoundland that are dominated by black spruce are referred to as a 'fire ecosystem,'—prone to fire because of less rain in the summer and a coarser soil. In fact, black spruce even requires the heat of a fire to open its cones and release its seeds.

In managing a forest, it is important to find a balance between suppressing unwanted fires and allowing a forest to take its natural course. Sometimes a fire is intentionally planned and set in order to prepare a logged area for reforestation. This is called a 'prescribed burn.'

Despite the fact that a fire can be important to the life of a forest, the reality is that the great majority of fires are caused by human carelessness, and are therefore not a natural occurrence. The Fire Weather Index (FWI) is a rating given to show the level of risk of forest fires for a given area—low, moderate, high and extreme. This is based on the relative humidity in the air and on wind factors. Campfires are prohibited when the FWI is "extreme."

The next time you are camping, check on the Fire Weather Index. You may have to do without your pot of beans over the fire, and settle for uncooked marshmallows!

169

Analysis:

5. *Review Figure 9.2. What would be the ideal age-class distribution of a forest in order to ensure a sustainable resource? Explain your answer.*

6. *Describe three things that trees compete for in the forest ecosystem.*

7. *How might the following silvicultural practices improve a woodlot: a. controlled burning; b. thinning and pruning; c. soil improvements?*

8. *Briefly describe three modified clearcutting methods. Suggest situations where each may be used.*

9. *Suggest two reasons why clearcutting is a more economical harvesting method than most.*

10. *Obtain a copy of the 20 Year Forestry Development Plan. Outline some of the present silviculture projects that will help forest renewal.*

Hemlock looper

Woolly adelgid

Spruce budworm

Another concern with herbicides is that they do not kill only one kind of plant—generally, all deciduous plants in the sprayed area will be killed. Inevitably this causes changes in wildlife populations. Many primary consumers (mainly insects) that were dependent on the plants killed by herbicides will die. This in turn will affect those secondary consumers (mainly insect-eating birds) that were dependent on the insects. However, this argument is not restricted to the use of herbicides. It applies to any events that cause significant ecological change.

Newfoundland Insect Infestations

Most insect populations tend to rise and fall dramatically, depending on a variety of factors. More often than not, pest populations are small, kept in check by a number of natural limiting factors including weather, parasites, predators and disease. However, when favourable factors combine, an outbreak can occur.

Two of Newfoundland's most economically valuable trees suffer from insect attacks—balsam fir and black spruce. While there are many insect pests in Newfoundland, three have caused the most damage.

Eastern hemlock looper: This insect is considered to be the most damaging. In its larval form, it feeds on both the old and new needles of the balsam fir, thus causing complete **defoliation** in one year. Between 1966 and 1971, the looper destroyed about 11.5 million cubic metres of economically valuable timber. As a result of this

170

outbreak, over 400,000 hectares of fir forest were sprayed with insecticide, thus saving an estimated 85 million cubic metres of wood.

Balsam woolly adelgid: This insect is a persistent pest that wedges itself under a tree's bud scales and its bark crevices. It is covered by a protective waxy coating that, along with its dwelling habits, makes it hard to control with insecticides. The woolly adelgid injects saliva into trees that deforms twigs and hinders bud growth. Over time, the top branches die off and eventually the whole tree might die.

Spruce budworm: In its caterpillar stage, the spruce budworm is a defoliator that feeds on new growth of fir trees. This activity can kill a tree over a number of years. Although the spruce budworm was a major problem in the mid 1970s and early 1980s, it is no longer considered to be significant (see spuce budworm case study).

Pest management has progressed significantly in recent years. Where aerial spraying with insecticides was once the only feasible method of control, other methods have recently been developed. The combination of practices used to manage forest pests is called *integrated pest management (IPM)*. IPM helps to reduce the vulnerability of the forest to pest damage while minimizing environmental side-effects. In this province, IPM is still in its infancy.

Foresters begin practicing IPM when they monitor many factors in the forests to determine if an outbreak is likely to occur. To assist in this process, Forestry Canada is developing computer models to help managers organize the vast amounts of information they collect and use it to predict possible outcomes. The factors considered in these models include: weather reports; continual measurements of pest population densities; defoliation damage patterns as observed by ground, aerial and satellite surveys; forest inventory information; and economic considerations. These models are known as *decision-support models* because they help foresters predict outbreaks and decide what to do about them.

If an infestation develops and some control is necessary, the forester has an increasing number of methods to choose from. Chemical insecticides, such as fenitrothion, have been the most common agents for control-

Defoliation: any action that causes the loss of leaves or needles. A pest that causes this type of damage is called a *defoliator*.

INTEGRATED PEST MANAGEMENT INVOLVES:

- monitoring systems
- rules for decision-making
- pest control methods, including pesticides, biological controls and other silvicultural methods.

WISDOM

There is a wisdom here
That humans need learn.
To live in one place
For a thousand years,
Maintaining balance with one's surroundings,
Remaining healthy,
Providing shelter, security, and nourishment
For other forms of life,
And when death finally comes,
Leaving the place not impoverished
But richer and gentler for your being there
Is a feat which has been beyond our grasp
Of any of human's ages,
Yet has been an integral part
Of this tree's consciousness
For millenia.

Tom Bender

...we first must have a biologically sustainable forest before we can have an economically sustainable yield (harvest) of any forest product...

C. Maser, 1990.

Fenitrothion

Fenitrothion is a chemical insecticide, approved for use by the federal government to control many insect pests, but there are still uncertainties about its effects. Commonly used in large-scale aerial spray programs, this chemical has been linked to lowering bee populations, which are important pollinators of flowering plants, including many agricultural crops. It may also be causing declines in forest birds, but no one is sure to what extent because bird counting methods are inaccurate. Nonetheless, unusual bird mortality has been detected in spray areas. A high proportion of birds in spray areas have also shown signs of a nerve sickness which has led to breeding failure. If insect-eating birds decline, there is a greater potential for future outbreaks. There is also concern about fenitrothion's impact on human health, aquatic invertebrates and fish, but so far there is little scientific evidence to support this concern.

Agriculture Canada is expected to conclude a review of the continued use of fenitrothion in 1993.

Adapted from *The State of Canada's Environment*, 1991.

ling infestations. Generally they provide good control, but can cause undesirable side effects. (see fenitrothion description in the margin.)

Biological pest control methods include the use of natural chemicals, disease agents, parasites and predators. The bacteria *Bacillus thuringiensis* or *B.t.* has become one of the most popular biological agents for controlling insects that experience a caterpillar stage. It is much less damaging to the environment than chemical insecticides, but it is more expensive and less effective when the weather is cold and wet. Unlike many chemicals that kill on contact, B.t. must be eaten by the insect to have an impact. It still cannot be considered harmless to the environment, because it can kill most moth and butterfly larvae, not just those that are causing the problem to the forest.

Pheromones are another type of biological agent. These chemicals occur naturally in insects, and are used to communicate with each other, often to attract the opposite sex. Specific pheromones can be used to attract and confuse a certain species of insect, thus interfering with its normal reproduction cycle. Pheromones are being used to monitor population trends of insects, but further research is needed to determine if they can provide effective pest control in Newfoundland.

Other biological controls that show some promise include fungi, viruses, parasitic worms, and a wasp that has helped control spruce budworm in Ontario.

Integrated pest management, which applies to agriculture as well as forestry, rarely attempts to eradicate a pest. Rather, it tries to keep pest populations at a low enough level to prevent an outbreak and significant forest damage. IPM involves using silvicultural techniques to prevent conditions that lead to an outbreak, and using chemicals and biological agents on a small scale at precisely the right time and place. Because of the economic and environmental costs associated with large-scale pest control, IPM is catching on fast as a preferred alternative to large-scale spraying of chemical insecticides. However, it is still a long way from being operational in Newfoundland and Labrador.

172

Analysis:

11. What are the three insect pests that have caused the most damage to Newfoundland forests? Research the life history of a pest other than the spruce budworm and describe how it damages the forest.

12. What is integrated pest management? Describe two components of integrated pest management.

13. What are some of the concerns that have been raised over the use of fenitrothion? Outline two alternatives to chemical spraying.

CASE STUDY
Spruce Budworm in Newfoundland

In the 1970s and early 1980s, Newfoundland experienced the worst outbreak of spruce budworm in its recorded history. The subsequent controversy surrounding the chemical spraying used to control the budworm eventually led to a Royal Commission on forest management in Newfoundland. The story of the budworm tells us a great deal about how destructive one pest can be and how difficult it is to find acceptable solutions to such problems.

The spruce budworm is the caterpillar of a small brown moth that feeds on the needles of balsam fir and certain species of spruce. It is always present in our forests, but usually at low population levels. An outbreak occurs when habitat and weather conditions combine to create a situation favourable to the budworm—large areas of even-aged mature and overmature trees, combined with warm, dry spring weather. Given these conditions, a small population can spread rapidly over a large area in two ways: the adult moths can fly or be carried by the wind; or the caterpillars, which spin a fine silken thread and dangle from the trees, can also be blown by the wind (see Figure 9.4 for life stages.)

Once the budworm finds a suitable tree, it begins to feed on the new growth. In this way, a budworm infestation can kill a tree in about three to five years. If the infestation does not last this long, defoliation damage will occur, but few trees will be killed. However, even if the tree is not killed, its growth is hindered and wood production is reduced.

Insect treatment programs are generally undertaken so as to protect the commercial value of the forest. If, however, we understand the forest first and foremost as an ecosystem, and only secondarily as a potential economic resource, we will be less likely to fall into the trap of declaring total war on all so-called pests.

Humber Environment
Action Group, 1992
*The Forests of Newfoundland:
An Alternative View*

Defoliation caused by the spruce budworm

173

A budworm outbreak can persist for 10 years, by which time the budworm will have 'eaten itself out of house and home' and other natural factors cause a decline. It is likely that this pattern of outbreak and decline has occurred for thousands of years, but it is only in the last few decades that the budworm's use of the forest has conflicted with that of humans.

The most recent outbreak began in 1971. By 1980, over one million hectares of forest were severely infested (see Figure 9.3). The dollar value of the dead and dying timber was estimated at $50 million in 1980. With these losses, and more anticipated, government mounted an immense spray program involving six different chemical insecticides as well as the biological agent, B.t. About $20 million was spent to protect the forest. The spray program triggered a public outcry over the safety of the chemicals being used and over forest management practices in general. The level of concern prompted the setup of the Royal Commission. The Commission concluded that, all things considered, the spray program was justified and that concerns for human health and the environment lacked sufficient evidence to support a decision not to spray.

It is interesting to note that one of the technical reports submitted to the Commission stated that large-scale clearcutting increases the susceptibility of a forest to future attacks.

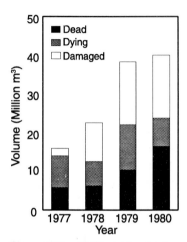

Figure 9.3 - Softwood volume of damaged, dead and dying trees caused by spruce budworm defoliation in Newfoundland, 1977-1980.

Imagine!

By 1980, about 400 common insect pest species were showing resistance to one or more pesticides.

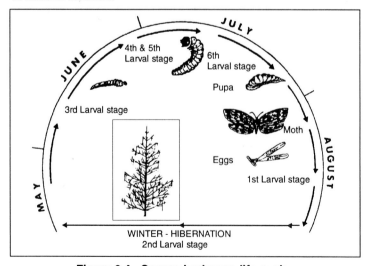

Figure 9.4 - Spruce budworm life cycle.

Last Thoughts

Forest management throughout Canada is facing difficult times as it shifts its approach in how our forests can be sustained. Sustainability once meant ensuring that an adequate timber supply would be available to serve future human needs. Now it means sustaining whole forest ecosystems as well as the economy. Many citizens are calling upon the country's forest managers to show a broader concern for all forest uses, not only those associated with commercial production.

Forest management is a tough job. It has to balance our immense economic reliance on forests with our need to keep them healthy. Any action that leads to a reduced harvest can mean a loss of jobs in the short term—but a sick and overused forest will be much more costly in the long term. Great improvements have been made in our use of the forests, but there is still much work to be done.

Environmentalists are often accused of painting an overly bleak picture of resource management practices. However, Forestry Canada commissioned an exhaustive survey of professional foresters in Canada (Omnifacts Research Ltd., 1991) and the results are truly alarming...
An astonishing 95% of Newfoundland foresters rate forest conditions as fair to poor, with major contributing factors including harvesting practices (cited as somewhat poor by 85%) and harvesting rate (82%).

Humber Environment
Action Group, 1992
The Forests of Newfoundland:
An Alternative View

Analysis:

14. *What triggers the outbreak of a spruce budworm infestation?*

15. *If you were the government minister responsible for forest management and your expert forestry staff told you that they expected another spruce budworm infestation, what would you do? Explain your answer.*

Students In Action – schoolgrounds silviculture

A woodland garden is a small-scale community of compatible shrubs, trees and other plants. As a class, you can choose a variety of seedlings and young native plants from the woods near your school. If you carefully transplant these, you can capture the sights, sounds and scents of a natural wooded area.

You don't need a large area, but you will need to make sure that the soil and light conditions are carefully evaluated to match the conditions required for your plants. Your regional department of forestry contact should be able to assist with this, and may also be able to help with the acquisition of some tree seedlings.

You will need to follow a number of silvicultural practices: planning, site preparation, planting, transplanting, tending, and protection. If you have a school compost heap, you can 'recycle' any weeds or cuttings from your garden, and use the compost to enrich the soil. Keep a record in your journal of the planning process, and ongoing observations of your woodland garden.

CHAPTER 10
Pulp and Paper Industry

With their pike-poles and peavies and bateaus and all,
And they're sure to drive out in the spring, that's the time;
With the caulks in their boots as they get on the logs
And it's hard to get over their time.

(From "The Badger Drive" by J.V. Devine. This traditional Newfoundland song depicts the life of the log drivers on the river, pushing the logs downstream to the mill. 'Pike-poles' ,'peavies', 'bateaus', and 'caulks in their boots' were all features or gear of the trade.)

Introduction

Who would think that such a common and useful substance as paper would have a high environmental cost? Unfortunately, it does. From the time a tree is cut until the paper is disposed of, we all pay a price for its use. In this information age, paper carries everything from world news to the words you are now reading. It provides the medium for both the lightest and most soul-searching thoughts of humankind. It packages many of the products we buy, providing protection and a means of promoting those products. It fills the aisles of supermarkets with products to keep us clean. And, more than any other substance, it is overloading our waste disposal sites.

Newfoundland and Labrador's extensive stands of black spruce and balsam fir have made this province a world-class paper producer. Pulp made from these trees makes some of the world's best quality paper because of its strength and ability to meet the demands of today's modern printing presses.

Paper production is a major contributor to this province's economy. In 1991, the value of pulp and paper production and the supporting forest industry was about $231 million. In total, the industry provided about 3600 **person-years** of employment.

About eight of every ten trees harvested in Newfoundland and Labrador are used to make newsprint. (The remaining 20% is used for lumber and firewood.) The newsprint producers are Corner Brook Pulp and Paper Limited in Corner Brook and Abitibi-Price in Grand Falls and Stephenville.

This chapter will look at the environmental issues surrounding the production of paper—with a focus on Corner Brook Pulp and Paper Limited—and the actions that both the company and government are taking to reduce the harmful environmental impacts of the pulp and paper industry.

What You Will Learn

- the importance of paper production to Newfoundland's economy;
- how paper is made;
- different methods of converting wood chips to pulp;
- the pulp and paper industry's effect on air and water quality;
- environmental regulations for the industry;
- the history and present status of Corner Brook Pulp and Paper Limited.

Imagine!
Paper was first made in China nearly 2,000 years ago.

Person-year: one person working for one year.

Analysis:

1. *What is the value of the pulp and paper industry's contribution to this province's economy?*

2. *Make a list of all the products around you that are made from paper.*

Paper Production
From Trees to Paper

Imagine!

The first paper mill in Canada was built in 1805 in St. Andrews, Quebec. The raw material used at that time was not wood, but rags.

This tree is almost as old as our country. Most of the events of our history could be ticked off on its crowded rings. The saw would have cut in only an inch or so before it passed my own birth year, and it would still have nearly a foot of wood to cross on the way to the pith. It gives one pause.
Gary Saunders
Alder Music: A Celebration of the Environment

In order to appreciate the environmental problems associated with paper manufacturing, it is important to understand how paper is made.

There are four main stages in the production of paper: removing the bark from tree logs; grinding the logs into chips; refining the chips into pulp; and converting this pulp into paper.

The bark is removed from the logs by rolling them around in revolving drums. These drums have a corrugated interior that tears the bark from the logs. The bark is diverted to furnaces where it is used to supply heat for other parts of the production process. The debarked logs are then moved onto a chipping machine that reduces the wood to pieces about 1 cm by 2 cm.

The conversion of wood chips to pulp involves breaking down the *lignin*, which is the natural glue that holds the wood fibres together. While the wood fibres are needed to make the paper, the lignin and other chemicals that are extracted or used in the process are discarded as waste products. The chips are converted to pulp in one of two ways: through a cooking and grinding process known as *thermomechanical processing* (TMP); or through a digester which uses a number of chemicals to turn the wood chips into pulp. A mill might use either of two chemical processes; kraft or sulphite. The kraft process, which can create a foul smell for up to 50 km from the mill site, is the dominant one used both in Canada and other countries.

In the TMP process, the chips are cooked in a vat of water at high temperature and pressure until they are soft. The vat is often heated by burning the bark removed from the logs. The steam created in this process is used to drive a variety of machines in other parts of the plant. The softened chips are then further refined into pulp in a number of grinding machines.

Many pulp mills throughout Canada, including Corner Brook Pulp and Paper, combine TMP with a sulphite digester process. The digester uses sulphurous acid to separate the wood fibres from the lignin and other impurities. In order to make the acid, sulphur is burned to produce sulphur dioxide, which is then passed through water to produce sulphurous acid.

The pulp from both the TMP and sulphite processes goes into storage tanks until it is required in one of the paper-making machines. Each type of pulp contributes different qualities to the paper, so they are usually combined to make the best product. Sodium hydrosulphide is also added to improve the brightness of the paper.

When the pulp (now called furnish) is needed at the paper machine, it is passed from the storage area to the 'head box'. The head box provides a constant, measured flow of pulp onto a fast-moving screen belt known as a *fourdrinier*. The fourdrinier moves the furnish through a series of hydrofoils and suction boxes that remove the water. As the moisture is removed, the wood fibres bond together to form a long sheet of paper. The paper is sent through rollers, pressers and dryers and finally collected on gigantic spools for sale and distribution.

Throughout the paper production process there are many kinds of waste produced. Two of the most common are emissions into the air and effluents into the water.

Analysis:

3. Briefly outline the paper-making process.

4. Distinguish between the TMP method and the sulphite or kraft method for converting wood chips to pulp. Which do you think is more damaging to the environment?

5. Identify some potential pollution problems in a pulp and paper mill operation.

Air Pollution

Before we look more closely at the environmental aspects of the pulp and paper industry, we must consider some characteristics of **air pollution** and how it is measured.

Air pollutants can be grouped as either *particulate* or *gaseous* matter. Particulate pollution refers to tiny, solid or liquid particles floating in the air. You can clearly see particulate matter in ordinary house dust if you concentrate on the air caught in the sunbeam of a window.

Particulate matter is measured primarily by the size of the particles. Since they are so small, these particles are measured in units called microns. (1000 microns equal

Air pollution: the transfer of harmful amounts of natural or synthetic materials into the air, usually as a result of human activities.

Substance	Max	Min
raindrop	5,000	500
tobacco smoke	15	0.01
carbon black	about 0.001	
fly ash	110	3
cement dust	150	10
house dust	200	20
plant spores	30	10
viruses	0.05	0.003
bacteria	30	0.2
gas molecule	about 0.001	

Table 10.1 - Examples of particulate dimensions, in microns.

The Influence of Moving Air

Air pollutants can travel thousands of kilometres on air currents to affect a part of the world far from their source. Air moves about in a cycle as it heats and cools. Air warms and rises around the equator. The turning of the earth forces the warm air toward the colder north and south regions where it cools, becomes heavier and falls towards the earth. The cool air then travels back along the surface of the earth toward the equator, warms there, and the cycle starts again.

one millimetre.) While particulates can range in size from 0.005 to 100 microns, fine particles—usually smaller than 2.5 microns—are considered the most hazardous. Generally, particles from 0.1 to 6,000 microns will settle to the ground, while smaller particles remain in the air. Table 10.1 shows examples of particulate dimensions.

Particulate matter is also measured in terms of how many are present in a quantity of air. The unit of measure is micrograms per cubic metre ($\mu g/m^3$). Picture the amount of paper underneath the period at the end of this sentence. That amount of paper weighs about 13 micrograms. Now think of the volume of a typical office desk. Most desks take up about one cubic metre. This should help you visualize one microgram per cubic metre, an important unit in measuring air pollution.

The two most significant air pollutants in the emissions from many pulp and paper mills are sulphur dioxide and particulate matter. The burning of the bark and the sulphite milling process have been the main sources of these pollutants.

Sulphur dioxide is a gas that occurs naturally in the atmosphere, but human activities can elevate it to unhealthy levels. The burning of coal is the greatest producer of this gas on a global scale, but combustion, such as bark burning at a paper mill, can be significant at a regional level. Sulphur dioxide is a major contributor to acid rain, and high levels of exposure can be harmful to both plants and humans.

Particulate matter can be a general nuisance when it accumulates on one's clothing or property, but it also contributes to lung disease and global warming.

The Newfoundland and Labrador acceptable air quality standards for particulate matter are 120 $\mu g/m^3$ average over 24 hours for short-term exposure and 70 $\mu g/m^3$ average over one year for long-term exposure. The law does not state the allowable size of particles, but particles smaller than 10 microns are considered more dangerous because they can penetrate deeper into the lungs.

A Closer Look: wood fibres

We are so used to the smooth, white sheets of paper we read from and write upon that it is hard to think of them being made of millions of microscopic ribbons. But they are. These ribbons were originally tiny tubes or fibres packed side by side in the trunks of trees.

Although in a cross-sectional view these fibres appear as open tubes, inside the wood they are closed-ended and very short—about .30 centimetres in length, and very thin—about .0025 centimetres in diameter. Separated from wood, they look like tiny needles.

To make paper from wood, it is necessary to separate the fibres from one another, treat them in certain ways so that the tubes collapse into ribbons, disperse them in water, and drain away the water on a fine screen. The fibres settle on the screen in a random fashion, crossing one another in every direction, but all settling flat on the screen.

No glue is needed to stick the fibres together because they all carry on their surfaces a natural adhesive. When the water is drained away and evaporated, this adhesive bonds them together in exactly the same way as a postage stamp is bonded to an envelope.

"From Trees to Paper"
Canadian Forestry Service fact sheet,1974

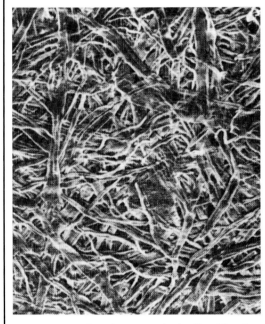

The surface of writing paper looks like this when magnified 220 times. Wood fibres, now flattened into tiny ribbons, overlap each other in all directions.

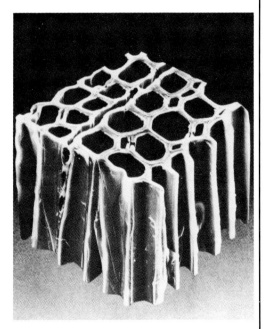

The tubular structure of wood fibres can be seen in this block of spruce wood, magnified 1,000 times.

Air Emission Standards

Air pollution is monitored and regulated by the provincial government through the Air Pollution Control Regulations. Table 10.2 shows the standards used to measure sulphur dioxide and particulates. The maximum acceptable standard shown in this table provides adequate protection against effects on soil, water, vegetation, materials, animals, visibility, personal comfort and well being. The maximum tolerable level indicates concentrations that require immediate reduction.

The Fish Test

One of the tests used to judge the level of pollution in effluent is to place rainbow trout in a sample flow of the water from a mill. If 20% of the fish die within a specific time (96-hour base period) the discharge is unacceptable and in violation of the legislation.

Chlorine: a naturally occurring element in the earth and seawater. It is most commonly found bonded with sodium as sodium chloride (NaCl), rock salt. In gaseous form it is highly toxic.

Dioxins and furans: popular names for complex organic compounds that are usually byproducts of chemical reactions involving high temperatures and chlorine. A few, not all, are highly toxic.

Water Pollution

Paper production uses vast amounts of water—about 100 cubic metres for each tonne of paper produced. This means there are also large amounts of liquid wastes or *effluents* which contain wood fibres, other fine solids, and chemicals. Scientists group these waste products into three categories: total suspended solids (TSS); waste which causes biochemical oxygen-demand (BOD); and toxic contaminants.

Total suspended solids are fine particles that will flow with a current but settle in slow or still water. In settling to the bottom, they coat whatever is there, blocking light from plants, smothering animal life and ruining bottom habitats such as spawning beds. TSS might also have toxic chemicals adhering to their surfaces, thus adding to their harmful effects.

Waste causing *biochemical oxygen-demand* refers to organic material that decomposes when released into the environment. This material serves as a vast food supply for microorganisms which use up oxygen as they carry out their decomposing function. When large quantities of material causing BOD are released, oxygen consumption can be so great that it is depleted from the surrounding water, killing fish and other aquatic life. The pulp and paper industry is a major source of material causing BOD.

Toxic contaminants include lignin, and can include many other chemicals that kill aquatic organisms. Three of the most noxious of these chemicals are **chlorine**, **dioxins** and **furans**, which are used in the bleaching process at many mills.

The standards for acceptable concentrations of TSS, BOD, and toxic contaminants in paper mill effluents are set by the Fisheries Act because of the potential for impact on aquatic environments. Known as the Pulp and Paper Effluent Regulations, they were last set down in 1971, but new legislation was passed in 1992. The changes require companies to comply with stricter standards than those previously in place (see margin note on page 184).

Air Pollutant	Max. desirable level in $\mu g/m^3$	Max. acceptable level in $\mu g/m^3$	Max. tolerable level in $\mu g/m^3$
Sulphur dioxide	450 per hour	900 per hour	X
Effect	Begins to damage vegetation	Odourous. Increasing damage to vegetation. Human health effects especially when linked with other pollutants.	Increasing health effects aggravation asthma and bronchitis.
Particulates	X	120 per 24 hours	400 per 24 hours
Effect	N/A	Reduced visibility. Soiling. Slightly reduced lung function in some children.	Increasing health effects aggravation asthma and bronchitis

$\mu g/m^3$ = micrograms per cubic metre, X = no value stated, NA = not applicable

Table 10.2 - Standards for measuring SO_2 and particulate matter.

Analysis:

6. *Describe three categories of waste products released in the effluent from a pulp and paper mill.*

7. *Name two toxic substances that are released from a pulp and paper mill.*

Corner Brook Pulp and Paper Limited

History

When the Newfoundland Paper & Power Company established a pulp mill at Corner Brook in 1925, it was considered to be the largest of its kind in the world. Ownership of the mill was later transferred to Bowaters Newfoundland Limited, one of a number of subsidiaries of the Bowaters economic empire based in England.

A bustling industry developed in Corner Brook as millions of logs were floated down the Humber River to the mill site. The grinders and sulphite mills churned the pulp into paper, which in turn fed the newsprint markets of the United States, the United Kingdom and other countries of the world.

By the late 1950s, however, forest fires—caused predominantly by coal-burning trains—had a severe impact on the availability of wood for the mill. In 1957, the mill itself was burned to the ground. The pulp and paper

Imagine!

The first pulp and paper mill in Newfoundland was built in 1867 at Black River in Placentia Bay.

Pulp and Paper Effluent Regulations 1992

The 1971 laws governing effluents from paper mills were replaced in 1992. The new regulations were established to take advantage of new technology available to control and monitor effluents. Under the new regulations:

- all mills are required to meet the same standards for effluent control which take into consideration the paper production rates of each mill;
- mills have to install up-to-date pollution control systems by December 31, 1993;
- mill operators have to constantly monitor and report on their effluents and provide opportunities for government officials to check on them;
- every three years mill operators have to study and report on the water systems into which their effluent is discharged in order to see if any fish habitat is harmed in any way;
- mill operators have to develop emergency response plans to cover the possibility of things going wrong.

In the short term, while the mills are being upgraded, the regulations are expected to create 17 to 18 thousand jobs across Canada. In the long term, they will create 1,000 permanent jobs.

industry of Corner Brook has faced difficult challenges since that time.

New federal regulations were established in 1971 to control the effluent discharged from pulp and paper plants. These new regulations were more lenient for existing mills than for newer mills. Older mills, like the one in Corner Brook, were not required to make immediate and costly changes to their equipment and milling processes that would substantially reduce the quantity or quality of its effluent. Nonetheless, the regulations gave the company a clear sign of the environmental standards of the future.

While the company knew it had to improve its effluent, another major problem was looming, not in the pulp and paper plant, but in the forests that supplied it. Insect infestations had begun to bear a major influence on the wood supply to the mill. Most notable among the insect pest problems was the 15-year spruce budworm outbreak which, according to the company, killed 30% of the balsam fir growing stock (see Chapter 9). Although insect-damaged wood can be used to make paper, it produces a lower quality of paper with more lint that does not hold ink as well as paper made from undamaged wood.

The high concentrations of insect-damaged wood within the company's leased land, together with the regulations governing air emissions and water effluents, and the international demand for high quality paper, placed Corner Brook's pulp and paper industry in a difficult position.

With considerable government incentives, the Bowaters mill was sold in 1984 to Kruger Inc. of Montreal in an effort to re-establish economic stability. It was felt that Kruger had the capital and the ability to make the mill viable. Kruger renamed the mill Corner Brook Pulp and Paper Limited (CBPP).

Current Status

With a production rate of 400,000 tonnes of newsprint per year, Corner Brook Pulp and Paper accounts for much of the direct and indirect employment in Corner Brook. For every job involved with the production of paper, there are two to three other jobs supported by it. Without the paper industry, the economic prospects for Corner Brook would be very bleak.

The newsprint manufactured in Corner Brook is sold in an international market, where there is very strong competition to supply high quality paper for to-day's high-speed and technically advanced printing presses. For example, the trend towards more coloured inks in newspapers has required paper producers to improve newsprint quality. The main competition for Newfoundland's pulp and paper industry comes from Scandinavia.

In order to supply high quality paper, Corner Brook Pulp and Paper must have a high quality wood supply that can be sustained for a long period. This allows the company to build trusting, long-term relationships with buyers. In order to meet world demand for modern newsprint, the mill operators must constantly upgrade their processes and equipment at considerable expense. In short, the cost of remaining competitive is very high. Added to this is the cost of maintaining essential silvicul-tural practices and controlling emissions and effluent.

In Chapter 9 you learned about the importance of silviculture in maintaining a constant, reliable wood sup-ply. With the aid of both federal and provincial govern-ments, CBPP's silvicultural activities, between 1976 and 1990, were valued at $27 million. Figure 10.1 shows the relative amounts of silviculture treatments used during this period. This level of silviculture will need to continue and even increase, because without it CBPP will run short of wood to supply its mill.

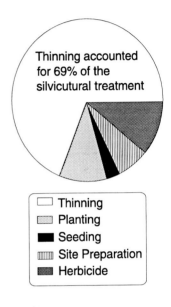

Thinning accounted for 69% of the silvicutural treatment

☐ Thinning
▨ Planting
■ Seeding
▥ Site Preparation
▦ Herbicide

Figure 10.1 - Silviculture treatments on CBPP land 1976-1990.

Problems with Air Emissions

The provincial Department of Environment and Lands monitors SO_2 and suspended particle levels in the air at Corner Brook and reports on them quarterly. Figure 10.2 shows the sulphur dioxide readings from the Depart-ment of Environment and Lands for December 1991.

In addition to monitoring sulphur dioxide levels, the provincial government also monitored particulate levels in Corner Brook using high volume air samplers at four locations. On thirteen occasions between March

...the environment is where we live; and 'development' is what we all do in attempting to improve our lot within that abode. The two are inseparable.

World Commission on Environment and Development

1990 and November 1991, the airborne particulates in Corner Brook exceeded acceptable limits. As a result of its findings, the Department of Environment and Lands informed the company of the problems.

When the company replied that the incidents were linked to problems at the acid plant supplying the sulphite mill, the Department recommended that the company take immediate preventive measures.

In a report dealing with the air emission problems in Corner Brook, the Department of Environment and Lands stated, "Occasional episodes of high particulate emissions from ...the pulp mill are inevitable with the current equipment..." The bark burners at CBPP were noted as substantial contributors to high particulate levels in downtown Corner Brook.

Because of the problems with air pollution, CBPP closed its bark burners and sulphite mill early in 1992.

Analysis:

8. *Refer to Figure 10.2 to determine those dates on which the hourly concentrations of SO$_2$ exceeded the limits.*

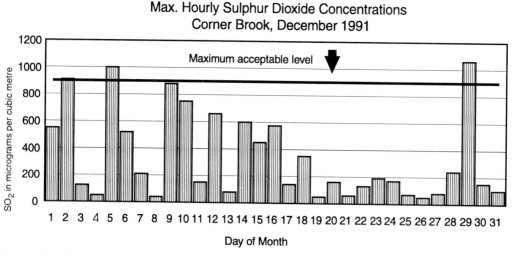

Figure 10.2 - Corner Brook air quality measurements (SO$_2$). Maximum hourly concentrations refer to the highest level of emissions produced in any one-hour period.

Problems with Water Effluent

CBPP is responsible for monitoring its effluent on a daily basis. The federal government can audit the company to determine if it is complying with the regulations, but the company is not required to make its figures available to the public. Therefore, no data were available on actual effluent readings from CBPP. However it is evident that the company is concerned about effluent levels since it is making significant improvements in the discharge of total suspended solids. Figure 10.3 shows the reductions in TSS that the Company reported for the period of January 1987 to January 1992. These reductions resulted from a number of improvements made to the company's milling procedures (see margin for details). In 1992, the company was also planning a system of reducing discharge by pumping air through the effluent and skimming off the sediment that rises to the top. All these changes were expected to lower the total suspended solids in the discharge by over 80%.

It should also be noted that Corner Brook Pulp and Paper uses sodium hydroxide, not chlorine, to brighten paper. This means that the mill's effluent does not contain chlorine, dioxins and furans.

Changes in Mill Operations

Because of problems with air and water pollution, and because of new demands in the newsprint market, CBPP made many changes to its operating procedures in the early 1990s. Some of these are described above and in "Reductions in TSS" in the margin. Other changes included the complete shutdown of the sulphite mill and a switch to thermomechanical processing, as well as a temporary halt to bark burning until a modern boiler system was installed. With U.S. newsprint markets demanding paper with significant recycled fibre content, a new plant was established in 1992, using 150 tonnes of previously used newsprint per day in the production process.

Imagine!

It is estimated that the rivers of the world transport one billion tons of solid debris to the oceans and about 400 million tons of dissolved matter.

Reductions in TSS

The reasons for the reductions in TSS as shown in Figure 10.3 were related to a number of specific improvements. In 1987, the Company switched to a dry debarking procedure. Prior to that, bark was removed with a water wash that discharged many of the bark particles with the effluent. In 1988, the milling process was upgraded to allow more water to be squeezed out of the pulp, and in 1990 the computer monitoring of the milling process was improved and fine-tuned to further reduce the suspended solids in the discharge. By late 1990 the filtering system had also been improved so that fewer solids were discharged.

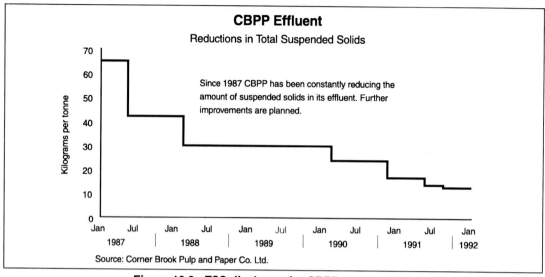

CBPP Effluent

Reductions in Total Suspended Solids

Since 1987 CBPP has been constantly reducing the amount of suspended solids in its effluent. Further improvements are planned.

Source: Corner Brook Pulp and Paper Co. Ltd.

Figure 10.3 - TSS discharge for CBPP, 1987 to 1992

Last Thoughts

Corner Brook Pulp and Paper Limited has continued to have difficulty complying with air emission and water effluent standards. The 1992 federal legislation governing water effluents made their situation even more challenging. The company faced a large expenditure for new machines and sophisticated computer technology in order meet both environmental and world market standards.

Through its Green Plan, the government of Canada is aiming to make this country one of the most environmentally-friendly nations of the world. To do this, it is clear that the pulp and paper industry must make some major and costly improvements in forest management and paper manufacturing operations. The difficulty arises when competing paper-producing countries are less concerned about the environmental impact of their industrial activities. Although all countries are selling their products on the same international market, some may be able to produce cheaper paper, and therefore be more competitive, if they are not required to invest in environmental controls. Clearly, it would be better if all

188

paper producers had the same environmental standards so they could operate on 'a level playing field'. Canadian companies could then play their part in protecting the environment without forsaking the communities that rely on their continued operation.

Analysis:

9. Describe two of the steps taken by CBPP to reduce pollution caused by its milling operation.

10. Do you think that governments should use tax dollars to help industry meet environmental standards?

11. Some North American industries have relocated to countries with lower environmental standards than Canada. How do you feel about this practice and what do you think of a company that would do this?

12. Why isn't more paper recycled?

13. What might be some environmental problems associated with paper recycling?

14. If more paper were recycled, might jobs be lost? What jobs might be created?

15. If environmental standards were less strict, perhaps industries could compete more easily and provide more jobs. Do you think environmental standards should be more lenient?

Students in Action – reducing the paper waste mountain

Paper is the leading contributor to our waste disposal sites. Most of the paper we use is thrown away again within a few weeks or, in the case of newspapers, within a day. We recycle less than 25% of the paper we use. If we recycled even 50% of the paper used in the world today, we'd save 10 million acres of forest from paper production. Here are some easy things you can do to cut down the amount of paper you throw away.

- Recycle! Cardboard, newspapers and non-glossy paper can all be recycled.

- Re-use! Keep a drawer for scrap paper. Use the blank side for doing rough work. Cut up scraps of paper to use for taking messages by the telephone. Save and re-use wrapping paper boxes and tissue paper. Re-use envelopes.

- Set up paper containers for your school and your home and see how much paper you can save from reaching the dump!

CHAPTER 11
Mineral Exploration and Development
Hope Brook Gold Mine

Let's face it, the southwest coast of Newfoundland is a difficult place to mine. It's much like tundra areas of the Arctic, where there are no trees to offer protection.

Bill Fotheringham, General Manager, Hope Brook Gold Inc.

Introduction

Minerals are essential to the functioning of modern societies. They are all around us, but only in a very few places do they become so concentrated that it is feasible to extract them from the earth's crust. Finding these concentrations of minerals, and then removing the precious compounds for our use, is not easy. We can pay high economic and environmental costs for their benefits.

This chapter is about one of the most precious and sought-after of all minerals—gold. We will look at the story behind one particular discovery of gold in the southwest corner of Newfoundland and the troublesome events that surrounded the early development of Hope Brook Gold Mine. As you study this chapter, remember that the events described here provide only one example of the interaction between a mine and the environment. Many mines go through a full cycle—from discovery to complete development of a mineral source—while causing only minimal impact on the environment.

Some Basic Geology

We tend to think of the earth as a pretty stable platform, but when viewed over millions of years, it is evident that its crust is constantly bulging, leaking, wrinkling, folding, sliding, cracking, and wearing down. (It makes you think twice about building a home here!) Although the earth's surface is quite active, it is only when an earthquake or volcanic eruption occurs that most people detect its instability. In order to understand why and where minerals become concentrated, we must first look at this changing earth.

The earth is a layered sphere. The super-hot inner core is molten, a metal soup that is composed mainly of iron and nickel. On top of this liquid is the *mantle*, a semi-solid and shifting layer. Although lighter and cooler than the core, it is still very hot. The closer the mantle is to the earth's surface, the cooler and stiffer it is. Floating on the mantle are *crustal plates*, which support the continents and form the basis of the earth's crust. This crust is made up of the lightest mineral compounds that

What You Will Learn

- the basic geology of mineral concentrations, especially gold;
- the need for an understanding of geology, geophysics, and geochemistry in exploring (prospecting) for gold and other mineral resources;
- techniques used in prospecting;
- background information on the Hope Brook Gold Mine in southwestern Newfoundland;
- mining methods and their environmental impacts at Hope Brook;
- the environmental consequences and options for dealing with the effluent wastes from the mine;
- the economic significance of the gold mine;
- the reasons for the Hope Brook Mine closure;
- the debate between mineral development and environmental protection.

Imagine!
Only a very small fraction of the area that is leased to mineral development companies ever gets converted to a mining operation.

Imagine!

Before gold production started, $200 million dollars was invested at Hope Brook.

Outcrop: bedrock that pokes out above the covering of soil, water or glacial deposits.

Geophysical anomaly: an unusual physical property in rock.

Metal anomaly: an unusually high concentration of metal in rock or soil.

Looking for Minerals

The work of mineral explorationists (prospectors) and geologists is detailed and painstaking. This means they must do a lot of homework before taking off for the wilderness. They start with maps that show the geology and topography (hills, plains and valleys) of a region. They look up records on previous exploration in the area. They conduct air surveys, using remote detection devices which search for **geophysical anomalies** and sample bedrock, stream and soil for geochemical analysis to detect **metal anomalies**. They accumulate data from known ore bodies to predict where new deposits might occur. All this information gives explorationists a good idea of where to look for valuable deposits.

have 'frozen' or crystallized to form rock. Convection currents in the core and the mantle provide the forces that cause the action in the earth's crust, including the slow shifting of the crustal plates.

Minerals are naturally occurring inorganic compounds that have formed from physical and chemical actions and reactions within the earth. As the earth evolves, the heavier metals such as gold are continually being moved from the mantle to the crust. Many geological processes cause these metals to become concentrated in large enough deposits to permit recovery by mining. Even after their formation, mineral deposits continue to be broken down, transformed or moved by weathering, sedimentation, leaching, heating, and other geological processes.

With knowledge of these processes, geologists can use the 'lay of the land' and **outcrops** to find clues about what lies beneath. The composition, structure, and shape of rock and its position on the earth's surface may reveal signs of economically significant minerals. Mineral deposits that can be profitably mined are called *ores*. Deposits with a high mineral concentration are *high-grade ore*, while those with low concentrations are called *low-grade ore*.

Finding Mineral Deposits

Although panning for gold in a cold mountain stream is still a valid technique used by some prospectors, significant mineral discoveries usually require far more sophisticated approaches. Today's prospector is likely to be a well educated scientist who uses electronic equipment to understand the earth's highly complex geology. Many are employed by exploration or mining companies.

Once mineral explorers have discovered an area with high mineral potential, they stake it out, register their claim with the provincial government, and take a closer look. There are a number of ways of doing this. *Geophysical* methods are those used to look at the physical properties of the rock. (Ore bodies have different physical properties than the barren rock surrounding them.) The geophysical methods include magnetic surveys, radiation detectors, gravity seismic readings and electromagnetic appraisal measurements. This part of exploration has little impact on the environment, since it uses sophisticated detectors that can sense the rock below.

Geochemical methods are used to look at the chemical composition of rock and sediment samples. Stream samples are especially helpful, because flowing water brings rocks and sediments together from miles around. Stream samples can reveal *placer deposits*, places where heavier metals or other minerals have settled out of running water as it slowed, while lighter debris was carried downstream. *Grab samples* are chipped off outcrops or from surface deposits and analyzed in a laboratory to determine what metals the rock contains. They often give some sign of what might lie deeper.

Roughly 95% of mineral exploration goes no further than these analyses. Even if a deposit is found, prospectors must consider the economic feasibility of extracting and marketing the deposit. It may be too small, too spread out or too hard to reach. The current and anticipated market price are major factors that can push an exploration forward or cause it to close down.

If the explorers are very thorough and very lucky, they might discover a significant mineral deposit. If so, the next phase of the work may have a more serious impact on the environment. Work crews, engineers and more geologists arrive to determine the size and location of a deposit. Heavy construction equipment or explosives are often used to create trenches that will reveal the length and width of a deposit at the surface. Drilling machinery is brought in to determine its depth. A road might even be built and a wilderness camp of considerable size might be erected. All-terrain vehicles are used to get people and equipment from place to place. All this activity and equipment can have a considerable impact on the environment.

In 1992, no environmental impact assessment was required for most mining exploration work, although certain specific activities were subject to environmental controls, such as altering the flow of a watercourse. Both within and outside the mining industry, many people feel that the environmental impacts of this stage of exploration are too slight to warrant many regulations or an environmental impact assessment.

Imagine!

In Newfoundland, 95% of the bedrock lies hidden by glacial deposits, soil, vegetation, lakes and bogs.

The Value of Mining and Exploration

The exploration activity that must precede any mineral production is worth quite a lot to this province. In 1990, it was assessed at about $12 million, and in 1988 at $41 million.

The value of mineral production in Newfoundland in 1990 was $883 million–$713 million from iron ore and $170 million from other products. Employment in mining in Newfoundland and Labrador during 1992 was expected to be 3600, down from approximately 4400 in 1990.

Imagine!

Mining pays well, accounting for more wage and salary earnings than the agriculture, forestry and fishing industries combined.

___

Imagine!

Geologists often collect plants rather than rock samples. Certain species of plants concentrate metal elements in their tissues. Above-average concentrations of a certain metal element in the plant tissue may be an indicator of a hidden ore body below.

___

Host rock: the rock within which a mineral deposit is found.

Some exploration activity can be disruptive and extensive, but this is not necessarily so. Some companies are making serious attempts to minimize the impacts of their exploration work.

When deciding whether or not to mine a deposit, one very important consideration is the concentration of the mineral in the **host rock**. Gold, for instance, rarely shows up as nuggets or obvious veins. It is usually sprinkled in microscopic flecks throughout the host rock. At Hope Brook, there is only about 4.44 grams of gold per tonne of host rock. This is why so much rock must be evaluated in order to determine the concentration of minerals in the ore body. This information helps the mining company decide how the ore should be mined and refined, and what it will cost to extract it.

If the work to this point suggests that there is a deposit worth extracting, the next step is to determine how this will be done and how the mineral can be transported to market.

Analysis:

1. *Research five metals that we use every day. Briefly summarize how they are extracted, and identify some of the potential environmental problems associated with this process.*

2. *"All industrial activity produces some effect on the biosphere of the earth, so the question of how much alteration to the environment is acceptable arises no matter what industry is involved. Mining is no exception." (The Northern Miner, 1990)*

   Interpret this quote, using an example that illustrates the importance of minerals to our society, while keeping in mind the consequences for the environment.

3. *Summarize the steps involved in mineral exploration in a flow chart format. Shade those steps in the process that might be damaging to the environment, and then identify some controls that might prevent this damage.*

4. *Through library research, find out more about the variety of remote sensing devices used in mineral exploration.*

5. *Contrast the development of nonrenewable resources (e.g. minerals) with the harvesting of renewable resources (e.g. fish). Give an example of how the former can affect the latter.*

Environmental Impacts of Mining

The impacts of mining can be significant. Although the industry has a poor record historically, modern mining has improved that record significantly. Yet, the industry is still characterized by problems that have no obvious long-term solutions (see margin).

One of the most significant potential impacts of mining is water pollution, which can arise from the removal of contaminated water from mines, from processing effluents, and from drainage of storage areas and abandoned sites (see Acid Mine Drainage in margin). During processing, the minerals are removed from the parent rock by use of strong chemical agents such as cyanide and sulphuric acid. Most modern mines are designed so that wastes are treated and stored on site. When waste material escapes into the surrounding environment, it is usually as a result of an accident.

Mining and mineral processing are regulated by a number of federal and provincial laws. The federal Fisheries Act requires that any industrial and municipal effluent entering a waterway be nontoxic. The Metal Minerals Liquid Effluent Regulations under the Fisheries Act establish the allowable monthly average and maximum concentrations of certain substances discharged by metal mines. Gold mining operations that use cyanidation (such as Hope Brook Gold Mine; see How the Mine Works, page 198) are exempt from these regulations, because at the time the regulations were being developed, no practical treatment method was available for dealing with cyanide in mining waste. Under the federal government's Green Plan, announced in 1990, new regulations and controls for metal mines are to be developed.

In Newfoundland and Labrador, industrial discharges are regulated by the provincial Environment and Lands Act. A mine operation in this province must have a certificate of approval stating the criteria to be met by the effluents, how they will be monitored, and other conditions.

Air quality controls for mining operations are governed by the Air Pollution Control Regulations of 1981 (see page 182-183, Chapter 10).

Potential Environmental Impacts of Mining

- major changes to the landscape
- road construction and loss of wilderness
- loss or movement of wildlife populations due to noise, human disturbance, and landscape alteration
- large piles of waste (tailings)
- water pollution from ore processing, sediment runoff, and drainage from tailings
- air pollution from milling, tailing piles and smelting
- high energy consumption
- collapse of underground tunnels
- inadequate reclamation after closure

Acid Mine Drainage

One of the most serious environmental problems with the mining industry in general is water pollution from acid mine drainage.

After the metals are extracted from the ore, the waste rock or tailings are discarded. Chemical and biological reactions within this discarded material can cause the production of sulphuric acid, which in turn begins to dissolve metals remaining in the waste rock or tailings. The acid and dissolved metals, both of which can be toxic, can percolate into surface and groundwater. In Newfoundland and Labrador, 5 of 91 mine sites were found to have potential problems with acid mine drainage in 1988.

Hope Brook

Origin of the Deposit

Millions of years ago, in what is now the Hope Brook area—5 km inland from Couteau Bay on Newfoundland's southwest coast —the earth's crust moved in response to mighty forces acting from below. Escaping, pressurized water was forced upward into the fractures of this fault zone. This water contained dissolved metals such as gold, silver and copper. As it rose into the mid levels of the crust, the metals precipitated out into the microscopic pores of the faulted rock, creating the gold deposit which is now being developed as a mine. Because this type of deposit was created from water and high temperatures, it is called a *hydrothermal deposit*.

Imagine!

Each U.S. space shuttle contains about 31 kg of gold.

A Closer Look: mining and wilderness.... compatible?

There is an ongoing debate between mining companies and land protection agencies. Each has a very different reason for wanting to stake claims to certain areas of the province. The group trying to preserve representative regions for future generations want those areas to be free from any resource development activity, including mining operations. The mining industry, on the other hand, fears that by setting aside land areas as parks or wilderness reserves, less land is available for mineral exploration, and in turn, limits are placed on opportunities for economic development.

At present, only about 2% of land in this province has protected status. Don Hustins, Director of Provincial Parks, argues that "the (parks) system has a potential for growing.... We hope that it will grow to be a bit larger. These lands represent the best of the natural environment, our own natural heritage. This is a program to protect our natural heritage before it's lost, destroyed, or developed."

Peter Dimmell, president of the Newfoundland and Labrador Explorationists, the province's main mining industry group, disagrees with the movement to protect land that excludes exploration. He claims that "the only way you can go in and enjoy those areas is if you have the leisure time and affluence to do it. And the only way we can have that is to have some sort of economic activity. If we limit the potential of economic activity, we also limit the use of these things."

(Quotes from "Miners Fight Newfoundland Parks," by Craig Westcott, in *The Northern Miner*, July 15, 1991.)

Think about these two positions. What other arguments could be made for or against establishing protected areas or mining developments? Is there room for compromise?

Recent History

Mining exploration activity in the area between Grand Bruit and Burgeo is not new. In 1902, the area was first claimed by John Chetwynd, a merchant in Grand Bruit who sank three exploratory mine shafts. Very little seems to have developed from this work, but the claim is still called the Chetwynd property.

In 1982, British Petroleum (BP) began working the area. In 1983, a pair of exploration geologists trudged northeast along the ridge behind Hope Brook. Ivor McWilliams spotted a highly mineralized zone and began to take grab samples. Eight weeks later, the results of the sample analysis showed that the rock contained ten grams of gold per tonne. (Three to seven grams per tonne were needed to make mining viable).

Within weeks, drilling equipment was airlifted to the site. Core samples revealed the continuing presence of gold to a depth of 300 m. Next, a massive helicopter brought in a backhoe to begin trenching. The operation grew steadily, until by 1985 the deposit had been fully mapped and measured. The operators estimated that about 10 million tonnes of ore would yield 38.3 million grams of gold over an 11-year period. Hope Brook Gold Inc., a subsidiary of BP Canada, was to be the sixth largest gold producer in Canada.

Before the mine could begin operations, the developers had to conduct an intensive environmental impact assessment involving both the federal and provincial governments. One of the tasks was to initiate a four-year study on the impact of the mine on the LaPoile caribou herd occupying the area. This research continued after the mine development began. After the assessment, the Department of Environment and Lands issued a certificate of approval which set limits on the concentrations of certain metals and chemicals in the mine effluent and stated that any discharge into natural waterways had to be nontoxic. This was the last step before the mine could begin to extract the gold ore for processing.

Hope Brook Ore
The ore at Hope brook is 94% silica (quartz) and 3% iron. The remaining 3% contains gold, copper, silver, lead, zinc and other trace elements.

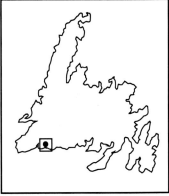

Map showing location of Hope Brook mine.

Imagine!
Political events in distant countries can influence mineral production in Newfoundland and Labrador. Internal political events in China during the late 1980s led that country to flood the world market with a number of minerals, thereby lowering their values. This caused the production of fluorspar at St. Lawrence to be no longer economically feasible, and in turn contributed to the closure of the mine there.

Construction of the heap leach pad and holding ponds, Hope Brook Mine, 1984

Canada is one of the world's principal exporters of minerals: 80% of domestic production is exported. The value of the export trade from the mining industry in 1990 was $18 billion.

Environment Canada

Gold is used:
• for health care
• in the aerospace industry
• in the electronics industry
• in jewelry, art and architecture
• as the world's monetary standard.

___

Imagine!

Ocean water contains an estimated 10 million tons of gold, but extraction is not economically feasible.

___

How the Mine Works

The upper 60 metres of the ore body was mined as an **open pit mine**. This gave access to the ore relatively quickly, thereby providing a source of revenue to the company while it developed the underground mine and the ore-processing facilities.

A process called *heap leaching* was used to extract the gold from the ore while the permanent ore-processing mill was being constructed. The heap leach process involved crushing the ore to pieces about one centimetre in size, mixing it with lime and piling the mixture on an impermeable sheet or liner. A weak cyanide solution was then sprayed over the crushed ore. The cyanide percolated into the fractures of the ore, dissolving exposed gold particles in a process known as *cyanidation* (see chemical equation below). The solution collected on the liner was drained into a lined storage pond.

The chemical equation for the cyanidation process is as follows:

$$4Au + 8NaCN + O_2 + 2H_2O \rightarrow 4NaAu(CN)_2 + 4NaOH$$

The solution containing the dissolved gold cyanide compound, now called the *pregnant* solution, was pumped through a series of steel tanks filled with activated carbon, similar to that in ordinary aquarium filters. The activated carbon adsorbed the gold cyanide compound from the pregnant solution. The remaining *barren* solution, now stripped of the gold cyanide compound, was refortified with more cyanide and recycled back to the heap leach pad. The gold cyanide compound was allowed to accumulate in the activated carbon until it reached the desired concentration and was later stripped from the carbon using a specialized electrical process.

Natural rainfall and snowmelt increased the volume of total solution above desired levels, so excess barren solution was occasionally drained off. This was treated with hydrogen peroxide to help break down the cyanide and discharged into the *tailings ponds*, a specially constructed holding area for the liquid waste.

The heap leach process is less efficient at extracting gold (64% recovery) than the more thorough milling process (87% recovery), because the cyanide cannot make contact with all the gold inside the coarsely crushed particles. This process was used for about a year, ending in 1988.

The permanent ore processing mill started operation in late 1988. In the mill, the ore is ground much finer than in the heap leach process and mixed with water to form a *slurry*. The slurry is pumped into tanks and mixed with a cyanide solution. The gold dissolves into solution as it did in the heap leach process. Activated carbon is then added to the slurry to adsorb the dissolved gold cyanide compound. Later, the gold-bearing carbon is filtered from the slurry by passing it through a screen. A complex process using a strong cyanide solution under high temperature and pressure then strips the gold cyanide compound from the carbon and puts it back in solution. This new cyanide/gold solution is passed through an *electrowinning cell*, where the gold becomes plated onto wire wool. The wire wool is then melted and the gold, being heavier than the other metals, sinks to the bottom of the molten bath, thus separating itself from the impurities. The gold is then poured into 32 kg bars and shipped to a refinery.

Imagine!
Gold makes up 0.00000035% of the earth's crust.

Two Mining Processes–why?
Two separate, but similar, ore processing methods were used at Hope Brook—heap leaching and a more thorough milling process. Both processes involve cyanide to leach gold from the ore. Heap leaching was used only for about a year to give the company some revenue while the mill was under construction. The milling process replaced heap leaching in 1988 because it is much more effective at removing gold from the ore. However, most of the environmental problems that arose were associated with the effluents from the milling process.

Canada lacks information about the extent and magnitude of environmental impacts caused by mining and processing of minerals; there are also insufficient data about industrial compliance with environmental regulations and guidelines...

Not only does Canada lack an accurate inventory of abandoned mine sites; the quantity of waste material stored in the country's estimated 6,000 abandoned tailing sites is unknown, and their potential hazard to the environment has not been assessed.

The State of Canada's Environment, 1991

199

Abandoned Mines
—Past and Present

Across Canada there are thousands of abandoned mines and tailings sites. Some of these cause serious environmental problems, while others simply limit the way in which the land can be reused. The regulations governing abandoned mines were much more lenient in the past than they are now.

When a mine is developed today, plans are made for what will happen when the mine closes. If no longer needed, buildings are removed, openings are sealed, and the site is restored as closely as possible to its state before the mine was developed.

In Canada, about 50% of the iron, 55% of the lead, and 40% of the copper produced comes from recycled material. Recycling of metals generated over $2 billion in 1989.

Environment Canada

At an earlier stage of this process—when the activated carbon was extracted from the slurry—the waste product, known as *tailings*, was treated initially with hydrogen peroxide, and later with sulphur dioxide. This was to neutralize the cyanide compounds in the tailings, which were then pumped to a series of tailings ponds. In the first pond, the solids settle out by gravity, while the liquids go on to a second pond where fine particles precipitate out of solution. The remaining solution goes to the *polishing pond*, where about 90% of the water is recycled back into the milling process, while the remainder is discharged into a natural waterway. High levels of precipitation prevent all of the water from the polishing pond from being recycled.

Problems at Hope Brook

The first gold was produced at Hope Brook in 1987, but by 1991 operations were suspended. A number of factors contributed to the closure, including a significant decline in market prices for gold, difficulties with the equipment used to haul the ore out of the mines and with the treatment of effluents. We will focus here on the problem with the effluents.

Effluents from both the heap leach and milling process were treated with hydrogen peroxide to convert cyanide into less toxic *cyanate*. This treatment had been used successfully elsewhere, and laboratory tests suggested that it would work for Hope Brook ore. During the heap leach phase of the project, effluents met the regulatory requirements. However, during the conventional milling process which followed, the hydrogen peroxide treatment of the tailings failed to remove or neutralize cyanide and copper adequately. (The leaching process removes copper as well as gold from the ore.) As a result of this failure, the tailings pond filled with solution that did not meet acceptable standards, causing a six-week suspension of operations in 1989.

The regulating agencies relaxed the effluent standards temporarily and allowed the mine to continue operating while new treatment technology was established. In 1990, the new treatment plant, using sulphur dioxide rather than hydrogen peroxide, was constructed. The new process significantly improved the discharge, but the tailings pond was full of contaminated solution from earlier discharges. Operations were suspended again in 1991 to treat the contents of the tailings pond.

Another incident occurred in 1989 when a valve malfunctioned, causing contaminated effluent to be discharged into Cinq Cerf Brook, a nearby salmon river. The contamination was brief, but it caused enough damage that BP Canada was fined $10,000. The company voluntarily agreed to bear responsibility for restocking the river with salmon.

The plummeting gold prices (from $560 in 1986 to $370 in 1991) plus legal and other major costs associated with correcting the engineering problems, caused the Hope Brook operation to look uneconomic for BP Canada. Hope Brook Gold Inc. went up for sale and 320 employees were laid off in July 1991. In April 1992, Royal Oak Mines Inc. of Vancouver bought the mine, and resumed operations in June 1992 with a smaller workforce of 240 employees. By early 1993, it was too early to say how the new owners will do in the long term, but so far they have worked hard to minimize environmental problems. With the help of the new effluent treatment plant, they are meeting all applicable environmental regulations, and outflow from the polishing pond is nontoxic.

Cinq Cerf Brook

Imagine!

Early studies at Hope Brook showed that areas up to 6 km from the mine site that were occupied by caribou before the mine construction were avoided by the caribou during construction. It is unknown to date if this avoidance will be lasting.

Analysis:

6. *Research one of the potential environmental problems related to mining. Describe the environmental impact in detail, and identify any actions that are being taken to overcome this problem.*

7. *Obtain copies of the following:*
   - *federal Fisheries Act*
   - *provincial Environment and Lands Act*
   - *provincial Pollution Control Regulations*

   Read the sections of these documents that pertain to mining activity, and decide whether you think the regulations are reasonable for all concerned.

8. *Outline the heap leach process used at the early stage of production at Hope Brook. Compare this method of gold removal with the milling process that replaced it.*

9. *What were two problems that eventually caused the Hope Brook mine to suspend operations in 1991?*

Daniel's Harbour —A Mining Success

Over 800,000 tonnes of zinc, valued at about $500 million, was mined at Daniel's Harbour on the Northern Peninsula from 1975 to 1990. During that time, Newfoundland Zinc Mines Inc. caused no serious environmental damage and it provided many jobs and other benefits to local residents. When the mine was closed, the company donated the office building to an arctic char aquaculture project, and a large storage shed became a recreation facility. Before the mine closed, work began on the restoration of the site to a stable and safe condition consistent with the surrounding landscape. The tailings area was seeded and fertilized, all openings to underground areas were sealed, and pits were graded and contoured so that the former mine site could be put to other uses. As of 1992, the reclamation of the site had cost about $1 million.

Last Thoughts

Mineral exploration and development contributes significantly to our economy and provides well-paying jobs to thousands of Newfoundlanders and Labradorians. Most people would agree that these benefits should continue, but not at too great a cost to the environment.

We have concentrated in this chapter on one mining project that has experienced environmental problems. It would be a mistake to think that all mining operations are like this one. For example, the zinc mine at Daniel's Harbour operated successfully for many years without major problems (see margin).

Mining and mineral processing can be environmentally harsh activities. However, the mining industry is showing a willingness to improve its environmental record. In 1991, it was estimated that the industry spent about 30% of its research budget in a search for environmental improvements. Meanwhile, governments are slowly improving their regulatory ability as well. As a result, a new trend is developing where companies and governments are working together in the search for solutions.

It is clear that in order to continue using the products of mining, we must minimize environmental damage wherever possible and make choices about how much damage is acceptable. In making these choices, we must be prepared to balance the long-term costs and benefits of mineral activity with the many priorities we hold as a human society.

Students in Action – don't take minerals for granite!

1. Consider some of the minerals you use every day. Where did they come from? Find out more about the processes used to extract these minerals.
2. Look at the recycling of minerals. Automobiles, appliances, building materials, batteries, and many more items contain metals that could be recycled.
3. Work with your local politicians to promote metal recycling, and to exclude sensitive natural areas from mineral development.

CHAPTER 12

Aquaculture

Aquaculture moves the fishery out of the costly and uncertain 'hunting and gathering' stage of development into that of residential cultivation of a food source.

From *Building on Our Strengths: Report of the Royal Commission on Employment and Unemployment,* 1986.

What You Will Learn:

- what is meant by the term 'artificial' environments;
- the definitions of 'aquaculture' and 'mariculture';
- comparisons between aquaculture and agriculture;
- the advantages of aquaculture over traditional methods of fishing;
- the different species of fish being produced in aquaculture operations in this province;
- the technologies and types of practices used in aquaculture operations;
- the differences between finfish and shellfish farming;
- ways in which aquaculture affects and is affected by human activity in an area;
- the economic significance of aquaculture to the province;
- the extent to which aquaculture is economically and environmentally viable in the province.

Aquaculture: the cultivation of aquatic plants or animals.

Main concerns in aquaculture:
- habitat control
- genetic manipulation
- economics
- disease prevention

Introduction

In order to improve our control of plant and animal production, we alter the environment in many ways. Agriculture is one of the more obvious forms of this type of alteration. A more extreme form occurs when we create entirely *artificial environments*, where most factors influencing the production of plants or animals are controlled. Greenhouses and aquariums are common examples of artificial environments, while the Biosphere 2 project described on page 27 is a more radical example.

If an aquarium is big enough and is used to produce large numbers of aquatic organisms, it then becomes part of **aquaculture**, a new and growing industry in this province. However, not all forms of aquaculture involve the same level of control as we would see in an aquarium. Some, like mussel farming, do little more than provide a place in which mussels can grow, while the natural environment provides everything else. This chapter will focus on aquaculture as it is practised in Newfoundland. We will look at its environmental implications, and the compromises that are occasionally involved with making the industry work.

Background

Aquaculture began a period of rapid growth shortly after the Second World War, around which time the world's consumption of fish grew from 20 million tonnes to over 65 million tonnes per year. It is predicted that by the year 2000, the world's population will devour 93 million tonnes of fish per year. Wild stocks are not expected to meet or sustain this level of demand. Thus the emphasis on aquaculture, which can not only help satisfy a growing appetite for fish, but can also provide important economic benefits to coastal areas. Many countries of the world are getting involved, but China, Japan, the Phillipines, Korea, Chile, Norway and Scotland are among the world leaders.

Within the field of aquaculture, the term *mariculture* is also used. It refers to the use of the marine or saltwater environment for aquaculture, as opposed to use of fresh-water environments.

Most aquaculture operations rely heavily on the natural environment. This is partly for economic reasons, since constructing a facility in which all environmental factors were controlled would be very costly. Luckily, nature provides large, inexpensive water environments for growing.

Imagine!

Aquaculture presently accounts for about 10% of global seafood production.

A Closer Look: hydroponics

An example of another artificial environment is a 'hydroponics' garden. The word comes from two Greek words: hydro, meaning water; and ponos, meaning labour. In other words, letting water do the work—no soil necessary!

Growing vegetables without soil may sound like a new idea, but in fact it has been around for a long time. Hydroponics was being practised thousands of years ago in the Hanging Gardens of Babylon. It was also used by the ancient Aztecs and the Chinese.

There is nothing magic about growing plants hydroponically. In conventional gardens, soil provides the necessary support for plants and root structures, as well as the source of nutrients for growth. In a hydroponics garden or greenhouse, these two functions are performed by another plant-support medium (e.g. gravel, sand, vermiculite, or floating plastic boards) along with a complete fertilizer solution added to the water. The fertilizer is mixed to provide the right amount of nitrogen, phosphorous, potassium and other nutrients required for the plants' growth. One of the advantages of this type of gardening is that vegetables can be available year-round, and because they are grown in a controlled environment, plants are also free of insects and insecticides.

Although hydroponics is a relatively new technology for Newfoundland, a large-scale project was developed here in the late 1980s—the "Sprung" hydroponic greenhouse in Mount Pearl, which grew mainly cucumbers. Unfortunately, this project proved to be too costly and not appropriate in size or scale for Newfoundland's environment. There have been other successful hydroponics initiatives since then, growing lettuce and other vegetables like tomatoes and cucumbers.

Analysis:

1. *List some biotic and abiotic factors that might be controlled in an artificial environment like an aquarium.*

2. *Why might some aquaculture operations benefit from warm water? From cold water?*

Shellfish: marine organisms that have shells, such as mussels, clams and scallops.

Finfish: typical fish species such as salmon, trout, cod and others.

Aquaculture cannot be practised just anywhere, however. For practical and economic reasons, the use of natural, sheltered coves and inlets makes good sense. Before selecting a site, factors such as weather, ice, water and air temperatures, salinity, water quality, and disease potential must all be taken into account. Some operations also need a cheap source of heat to warm large quantities of water. Waste heat from nuclear and oil-fired power plants can provide this in some cases.

Aquaculture is similar to agriculture in many ways. In fact, a fish farmer and a grain or produce farmer have much in common. Both must thoroughly understand the conditions of the environment they are working in, and they must choose their sites and species to fit these conditions. Both also rely heavily on the science of genetics to produce hardy and productive crops.

Shellfish		Finfish	
Problem	**Solution**	**Problem**	**Solution**
Animals kept in close confinement can be susceptible to disease. Diseases can spread to wild stock.	Control potential disease problems with good management.	Animals kept in close confinement can be susceptible to disease. Diseases can spread to wild stock.	Control potential disease problems with good manage-ment, close veterinary attention and drugs.
Feces accumulate below lines, altering bottom conditions.	Choose site with good water circulation to help flushing.	Uneaten food and feces accumulate below cages, disrupting bottom life and enriching surrounding water with nitrates and phosphates.	Control feeding; rotate farm site; choose site with good water circulation to help flushing.
Waterways blocked by equipment.	Plan position of equipment to allow access.	Waterways blocked by equipment, interfering with commercial fishery and other boaters.	Plan position of equipment to allow access.
Unsightly equipment hurts tourism.	Avoid high tourist areas; avoid highly scenic areas; interpret the facility as a tourist attraction.	Unsightly equipment hurts tourism.	Avoid high tourist areas; avoid highly scenic areas; interpret the facility as a tourist attraction.
Introduction of new species or strains.	Control through regulations.	Introduction of new species or strains.	Control through regulations.
Risk to consumers from contaminated shellfish (e.g. PSP, industrial pollutants).	Test shellfish and close areas when contaminants are found; stop raw sewage dumping and reduce agricultural runoff which contribute to algal blooms; choose sites away from industrial pollution.	Drugs used to control disease and promote growth.	Control drug use based on veterinary advice; stop drug use for appropriate period before harvesting.
		Escaped fish might interbreed with wild stocks.	Make fish sterile so they cannot reproduce or inter-breed.
Occupy areas of seaduck feeding habitat.	Restrict area used for aquaculture and avoid known seaduck feeding areas.	Disposal of dead fish.	Recycle as fertilizer.

Table 12.1 - Some environmental problems and solutions in aquaculture.

It is the genetic manipulation of captive fish stocks that poses one of the greatest environmental concerns with aquaculture. It is assumed that non-native or genetically changed species will escape into the wild at some stage. The result would be mixing and interbreeding with wild populations, which could prove to be very damaging (see page 66-67 for a review of the potential dangers of introducing new species into ecosystems). Table 12.1 provides a summary of other environmental problems and possible solutions associated with aquaculture.

Analysis:

3. What might have been the reason for a growth in aquaculture after the Second World War?

4. In table format, compare aquaculture with agriculture. In what ways are these two activities similar and dissimilar?

5. Technology is playing an increasing role in the development of modern aquaculture. Research one example where technology is being applied.

Aquaculture in Newfoundland

The first serious attempts at aquaculture in this province began in the 1960s, with early research focused on the feasibility of mussel and scallop farming. By 1974, the Upper Trinity South Development Association was active in trout farming. But it was not until 1984 that the provincial government officially recognized aquaculture as an industry worth developing in the province. The interest seems to have developed when a Newfoundland delegation was studying Norway's offshore oil production and coincidentally noticed its flourishing aquaculture industry. Considering the similarity of coastal conditions in the two regions, it looked like a feasible industry for Newfoundland as well.

By the end of 1984, research into mussel farming had increased and the provincial government had established a salmon aquaculture research facility at Bay d'Espoir. When the Mussel Culture Incentive Program was developed in 1985, mussel culture kits were distributed by the government to potential farmers.

Imagine!

Besides finfish and shellfish aquaculture, seaweed and crustaceans such as shrimp are also cultivated. In fact, seaweed and shrimp culture are more important globally than Atlantic salmon culture.

Intensive versus Extensive Aquaculture

Intensive aquaculture refers to situations where a high degree of control and feeding is used to grow the product, such as with salmon or trout. *Extensive* aquaculture refers to situations where no purposeful feeding is done, such as with mussel or scallop production. Around the world, the older, extensive aquaculture is more important than the newer intensive form.

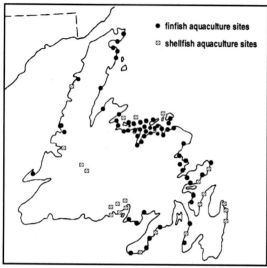

Figure 12.1 - Aquaculture sites in Newfoundland. Some sites are experimental and may not continue operation.

Table 12.2 shows the scale of the aquaculture industry in Newfoundland by 1991. At that time, the industry was employing about 250 workers. Although the record looked encouraging, some experts considered that most aquaculture operations in Newfoundland were still in the experimental or early production stages at that time. Few were fully proven as viable commercial enterprises.

At this stage in its development, Newfoundland's aquaculture industry is strongly supported by government. The provincial government issues aquaculture licences and provides funds, training and research. The federal government provides funding, research and monitoring of aquacultural sites. Federal and provincial experts also work closely with private entrepreneurs to develop and apply many of the specialized techniques required in this field. Potential aquaculture operators must meet many municipal, provincial and federal regulations before obtaining a licence. Environmental controls in aquaculture are quite strict and apply to all operators involved. Figure 12.1 shows active aquaculture sites in Newfoundland in 1992.

One of the most important concerns in the industry is the need to develop competent operators who are well trained in both the technological and business aspects of the industry. Because of this, aquaculture is gaining con-

Species	# of Sites in 1991	1991 Production in tonnes	$ Value in thousands
Mussels	60	320	$546.5
Scallops	13	2	$8.5
Trout	12	66	$247.5
Salmon	4	66	$260.0
Char	7	0.5	$2.1
Cod	4	80	$64.0
Ocean Pout	1	X	X
Lumpfish	1	X	X
Wolffish	1	X	X
Total	103	535	$1,128.6

Table 12.2 - Species grown in Newfoundland an Labrador aquaculture.

siderable attention in Newfoundland's education system. Courses are now offered at Memorial University and the Marine Institute, as well as a number of regional colleges.

Newfoundland aquaculture is divided into two main branches; *finfish aquaculture* and *shellfish aquaculture*. Finfish aquaculture refers to the production of species like trout, char, salmon, cod, and others, while shellfish aquaculture involves species like mussels and scallops. Many of the environmental concerns surrounding the industry relate to finfish rather than shellfish farming.

In 1992, shellfish farming—mainly of blue mussels—was the most economically significant form of aquaculture in Newfoundland. It also had the most sites and showed the greatest potential for growth. Studies suggest that Newfoundland's waters are capable of producing 5,000 tonnes of mussels per year—4,680 tonnes more than were produced in 1991. This level of production would make Newfoundland the largest producer of cultured mussels in North America.

Finfish farming in this province involves the cultivation mainly of **salmonids**, although other species are also being farmed (see Table 12.2). One of the main salmonid production sites is located at Bay d'Espoir, where the goal is to increase the 1991 production rate of 66 tonnes to 450 tonnes in the future.

Some finfish operations practise *fish ranching* rather than *fish farming*. In a ranching operation, the fish are released to the wild after hatching and rearing. Due to their anadromous nature and careful genetic selection, the fish will return to the hatchery for harvesting. In a fish farming operation, the fish never leave captivity.

Salmonid ranching is not practised in Newfoundland, mainly because of the need to prevent native stocks from interbreeding with genetically altered cultured stock. Federal scientists are concerned that if captive raised stock were to interbreed with wild stock, the genetic makeup of the offspring might be so disturbed that they would become lost and never return to their spawning grounds. It is not yet clear whether a fish's ability to find its spawning area is connected with genetics, but the risks are considered to be too great to take the chance. Another concern is that captive stocks can either intro-

Salmonid: member of the salmon family, which includes salmon, trout and char.

Wolffish Aquaculture

Experiments were underway in 1992 on wolffish farming. The blood of this fish is very valuable to medical science. It contains a natural 'antifreeze' that is used to preserve human organs being transported for transplant. Demand from the United States for wolffish blood far exceeds current production.

Source of Salmon Stocks for Aquaculture

Atlantic salmon from Newfoundland are less desirable in the local aquaculture industry than those from New Brunswick. During experiments done at Bay d'Espoir on the Grey River, Conne River and Exploits River stocks, they found that Newfoundland salmon grow slowly, mature too quickly, require more food per gram of fish produced, and have inferior flesh quality, as compared with those from the Saint John River, N.B. However, research continues in an effort to find a Newfoundland source of salmon eggs that will perform well in the aquaculture industry.

Steelhead trout: a kind of rainbow trout that spends part of its life at sea, like salmon.

duce new disease or, when interbred, make existing stocks more susceptible to disease.

Salmonids in Bay d'Espoir

A company called S.C.B. Fisheries in Bay d'Espoir is active in a variety of aquacultural activities, one of which is salmon production. The company's aim is to develop a commercially competitive operation using Atlantic salmon—a goal not quite reached in 1992, since many of the techniques and methods of the operation were still in development. In order to raise the money needed for the salmon operation, the company was producing **steelhead trout**, a species for which most of the technology and expertise was available.

Analysis:

6. *Distinguish between finfish and shellfish aquaculture under the following headings: present economic potential for the province; and the environmental concerns associated with each.*

Chromosome: that part of a cell that contains the genetic code.

The Switch from Freshwater to Saltwater

Salmon start life in freshwater and remain there until a dramatic change occurs within their bodies, allowing them to tolerate saltwater. At this stage they are known as smolts and are ready to migrate to the sea. The salmon farmer must duplicate this natural process by moving the the fish from fresh to saltwater. At this time they can tolerate full strength saltwater, but at Bay d'Espoir the best results are achieved in water that is only partially salt, or brackish.

Atlantic Salmon and Steelhead Trout

Salmonid farming is a complex business. Both steelhead trout and salmon develop in freshwater and then move out into saltwater. In each environment, they have specific temperature, salinity and food requirements for different stages of growth. In aquaculture, these needs are met through the careful selection of sites and through the control and monitoring of these conditions.

Because of the concern about captive stocks escaping and establishing new 'foreign' populations, or interbreeding with wild stocks, considerable work has been done to minimize this possibility. The aquaculture industry and government have been working together to develop captive stocks that have no chance of interbreeding with wild fish or with each other.

The process involves changing the fish genetically to make them *sterile*, unable to breed. It is a complicated process, but one that involves two major steps. To understand it, it is necessary to quickly review how the two sexes are determined.

The sex of a fish is determined by a combination of two **chromosomes**, X and Y chromosomes. Male cells have one X and one Y chromosome, while female cells have two X chromosomes. When sperm is produced in the male fish's testes, cells divide to create sperm with one X chromosome or one Y chromosome. The eggs produced by the female's ovary will all contain one X chromosome each. If a sperm carrying an X chromosome fertilizes an egg, the resulting offspring would be a female fish (two X chromosomes). If the sperm carries a Y chromosome, the offspring will be male (one X and one Y chromosome).

Back at the aquaculturalist's laboratory: the first step involves treating young fish with the male **hormone**, *methyltestosterone*, a drug that will make genetically female fish (having two X chromosomes) develop testes and produce sperm. This treatment is a sort of sex change operation that makes the fish look and act like a male, although the fish remains a female at the cellular level. The sperm produced by this customized fish all bear X chromosomes; no Y chromosomes such as a typical male would have. Next, the sperm from the sex-changed females are used to fertilize the eggs of normal females which also bear only X chromosomes. This process creates all female offspring with XX chromosomes. Up to this point, the fish are still capable of reproducing with wild males, so another major step is required.

The eggs go into a pressurized vat where they receive a shock that forces the eggs to develop a third chromosome. After this treatment, the fish are now genetically XXX and are known as *triploid* stocks, in contrast to normal fish which are *diploid* (having two chromosomes). This process prevents the fish from reaching sexual maturity, and it also prevents them from producing viable eggs. From the point of view of the aquaculturalist and the consumer, these fish have other advantages as well. With reproduction made impossible, practically all energy goes into growth—and they grow quickly.

After these specialized treatments occur at the hatchery—the only true artificial environment in the entire operation—the young fish are fed and moved from one type of water environment to another, depending on their stage of development. They remain at the hatchery for many months before moving to a saltier and colder environment. At the Bay d'Espoir hatchery, they have the advantage of waste heat produced by the hydroelectric

Hormones: naturally produced chemicals that control the functions of many organs. Hormones produced by the testes and ovaries help create the male and female characteristics of animals.

On Being a Finfish Aquaculturalist

Each step of the finfish aquaculture process requires a thorough understanding of fish biology and the aquatic environment. Even though the fish are held for the most part in cages submersed in a natural aquatic environment, the choice of the site and the species that can grow best in there requires a broad base of knowledge and training.

Environmental Concerns in Freshwater

All salmonids are raised at least partly in freshwater. During this phase, precautions have to be taken to minimize pollution of natural waterways. Effluents must be controlled so as to prevent contamination and excessive material causing BOD (see page 182 for a review of the importance of BOD).

Brackish: mixed fresh and salt water.

facility. The selection of sites with the correct environmental conditions is critical for the next stages of development. For example, the right level of salinity is very important—the ideal is 17 parts of salt per 1000 parts of water. At the Bay d'Espoir facility, the correct salinity range has been found at Roti Bay, where the water is **brackish**.

Protection from winter ice is also important. Roti Bay has no arctic or drift ice, but it can freeze to a depth of one metre. Fortunately, the fish can sink below this depth and still be in suitable conditions of temperature and salinity.

The final stage is probably the simplest to manage. The fish are simply held in saltwater cages where they are fed until they are big enough to harvest. At Bay d'Espoir, the salmon are held in Gaultois Passage, and the steelhead trout at Lou Cove.

Analysis:

7. Genetic engineering plays a big role in finfish aquaculture, as well as in the development of disease-resistant plants and plants that produce a higher yield. What is your view on genetic engineering? Is there a point where we should stop, or do we have the right to continue modifying organisms to meet our needs?

8. How do biologists prevent cultured fish from altering natural fish stocks?

9. If there is an aquaculture operation near your community, arrange an interview with the operator to find out how that particular operation works.

Blue Mussels

Imagine!

Mussels are the most numerous of all shellfish. They are hermaphrodites—having both male and female organs. They incubate a jelly-like spawn and then release into the sea huge numbers of barely visible young fry that float freely before eventually securing themselves to solid objects such as rocks or wharves.

Mussel farming is very similar to agriculture, with coves and inlets serving as 'fields' for mussel production. In these watery fields, the crops of mussels grow on floating lines called *longlines* (see Figure 12.2).

The longline method of mussel farming consists of a long rope floating near the surface of the water and anchored on either end. Suspended from it are shorter lengths of rope during the first stage of production. In the second stage of production, these are replaced by *socks*—fine cylindrical netting material that are much like nylon stockings.

The best sites for mussel production are bodies of water about 15 m deep or more, with the right salinity and temperature conditions as well as good tidal circu-

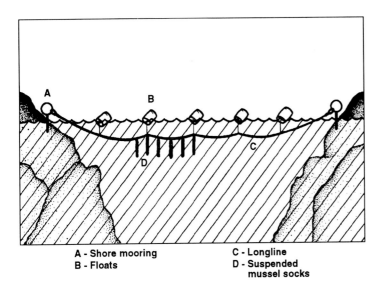

A - Shore mooring   C - Longline
B - Floats   D - Suspended
                        mussel socks

Figure 12.2 - Mussel longline.

Mussel Mud and Overgrazing

Beneath the longlines of cultured mussels, feces slowly accumulate on the bottom. This 'mussel mud' is a kind of pollution that can alter the natural community beneath the lines.

Overgrazing refers to the intense mussel feeding that takes place around growing operations. The mussels can devour the phytoplankton in an area, making less available to other species.

lation. They should also be near abundant wild mussel beds, which provide spawn to the farm operation. Since these sites must also be protected from extreme weather, small coves and inlets are ideal.

The first stage of production involves the collection of *spat*—baby mussels that grow from drifting mussel larvae (see Figure 12.3). These larvae are the products of mussel spawning, which occurs between June and September. The larvae will drift about and cling to any surface that seems appropriate, such as the short lengths of rope hanging from the longline. After about a year, the ropes are gathered in and stripped of their spat.

Roughly 800 to 1000 of the spat are placed into the nylon socks and attached again to the longline. The mesh of these socks is wide enough that the spat can work through the fabric and attach themselves to the outside of it. The socks are suspended again from the longline. They will remain there for 24 to 36 months, or until the mussels are large enough to harvest—about 50 mm. Mussel farmers can harvest about five to six kilograms of mussels for each metre of sock used.

The mussels are harvested by pulling the socks, which can weigh up to 30 kg, into a boat and transporting them to a processing plant. There they are removed from the socks, declumped, sorted by size, and stripped of the

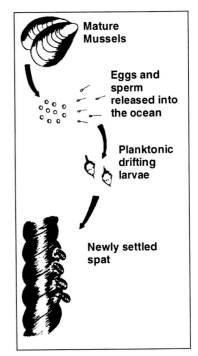

Mature Mussels

Eggs and sperm released into the ocean

Planktonic drifting larvae

Newly settled spat

Figure 12.3 - Life cycle of the mussel.

Diarrhetic shellfish poisoning: a type of poisoning in humans that causes diarrhea.

Paralytic Shellfish Poisoning

PSP results from eating shellfish that contain the spores of dynoflagellates. Of the 1200 kinds of dynoflagellates, only a few produce the poisonous substance known as *saxitoxin*. If the organisms producing the toxin are present in large numbers, the shellfish will ingest the toxin and it becomes concentrated in their tissues. In mild cases, nausea and vomiting might occur, while in more severe cases, paralysis of the limbs can develop and death may occur from respiratory failure.

An area can become contaminated when there is the right combination of temperature and nutrients in the water, allowing the organisms to multiply rapidly. This is known as a *bloom*, which will usually last for a few weeks. PSP occurs in all coastal regions of the world, including our own.

hairs that they use to attach themselves to the bottom. At that point, they are ready for market.

Although this process sounds relatively easy, mussel farmers face a number of difficulties. Crabs and starfish like to eat mussels, so mussel farmers have to constantly remove these predators from the longlines. Eider ducks can also be attracted to the abundant mussel supplies. Other problems arise with microscopic organisms. Because the mussel is a filter feeder, it sucks in water and absorbs whatever is contained in that water. Some of the organisms that it ingests can be harmful to the people who eat the mussels. For example, *coliform bacteria*, which are abundant in animal excrement, can end up in mussels that live close to a location where raw sewage is dumped into the sea.

Another common organism that causes problems for mussel farmers is a type of algae known as a dynoflagellate, which can cause *paralytic shellfish poisoning* or *PSP* (see more on PSP in the margin). Dinophysis, another microscopic organism, causes **diarrhetic shellfish poisoning** (DSP) in humans. It has recently been discovered in Newfoundland waters in minute quantities, but is much more widespread throughout the Maritimes.

Unfortunately, there are no controls for these problems. Obviously you would be well advised not to set up a mussel farm near a sewage drain, but there is no sure method of avoidance or control for other natural pollutants. DFO tests the mussels to see if any of these problems are occurring. If they are, the mussels in an area may be declared unfit for consumption. This can be disastrous for many mussel farmers, because people will avoid buying mussels from a large area even if the problem exists in only a small area.

Activity: Growing Fish—Bigger and Faster

A number of different factors influence the growth rate of fish. Food is one; temperature is another. Try an easy experiment by setting up two aquaria, each with one fish in it (goldfish will work fine). Keep one aquarium warm, about 30° C, by putting it in a warm place or installing an aquarium heater. Keep the other relatively cold, about 20° C. Feed both fish the exact same amount of food each day. Which fish is larger at the end of the month? What are the implications of this for an aquaculturalist?

Last Thoughts

It would be a mistake at this stage (in 1993) to think of aquaculture as a major economic salvation for this province or as something that can replace the commercial fishery. Its potential economic benefits appear to be limited, especially by the ice conditions that surround our coast. It is an industry that requires considerable expertise and sophisticated techniques in genetics and fish husbandry. It also requires good business and marketing skill to make it financially viable. With this needed commitment of knowledge and skill, however, there is little doubt that Newfoundland can operate a sustainable aquaculture industry—one that can add important economic diversity to our coastal communities and fine food products for the world's dinner tables.

Although certain environmental concerns will continue to surround the aquaculture industry, these appear to be relatively minor in light of the potential benefits. The principles of sustainable development—where resources are managed so that future generations will have the same opportunity to benefit from them—seem to apply well to the aquaculture industry.

Private vs. Public Property

One of the controversies of fish ranching concerns the public loss of access to certain areas as a result of their being turned over to private or commercial operators.

Analysis:

10. Study a map of Newfoundland and Labrador. Based on the characteristics of the coastlines, what areas of the province seem to hold potential for shellfish farming?

11. What are two other factors that have to be taken into consideration when selecting a site for shellfish farming?

Students in Action – buy local!

One contribution you can make to the development of aquaculture and other local industries is to buy local goods. Check the labels on fish and shellfish products to see if they were grown and processed in Newfoundland and Labrador. Insist on quality, but consider purchasing goods that are made here, even if it means paying slightly more. If enough of us buy these products, the prices will eventually come down and the quality will improve. You'll be supporting our people and our future!

TROPICAL RAIN FOREST DEPLETION

ACID PRECIPITATION

OZONE DEPLETION

UNIT 3
Global Environmental Issues

GLOBAL WARMING

OCEAN POLLUTION

THE SEARCH FOR AND USE OF ENERGY

HAZARDOUS WASTES

Unit 3

This next unit explores some of the most serious threats to the health of our planet today. Consider the magnitude of the problems of rain forest destruction, hazardous waste disposal, ocean pollution and acid precipitation. Even though many of these issues may seem far removed from Newfoundland and Labrador, we are still part of these problems and we have an obligation to work alongside the rest of the world to discover solutions.

If each of us starts to 'think globally and act locally' we can take responsibility and control for our part of the planet. The decisions and actions we take here at home can work towards moving our community, our society, and our world in a particular direction. We are responsible, and we also have the power to act.

Your individual actions matter! Together, the actions of many individuals can add up to substantial change and create even greater hope for the future of our planet and our children. In the words of local songwriter, Eric West, "we can all make a difference, you and I."

What You Will Learn:

- an awareness of the major global environmental problems, the basic science behind them, their current status, and any progress made toward the solution of these problems;
- the global connections between development and the environment, and a review of the concept of sustainable development;
- the special problems of developing countries in dealing with environmental issues;
- the connection between the global situation and our own local environmental issues;
- an interpretation, analysis and evaluation of information related to global environmental issues;
- a sense of personal responsibility and empowerment in relation to environmental problems—'thinking globally, acting locally';
- an understanding that individual actions have a cumulative effect and can make a real difference to environmental issues and their solutions.

CHAPTER 13
Tropical Rain Forest Depletion

If a tree falls in the forest, does anybody hear? Anybody hear the forest fall?

(From the song, *If a Tree Falls*, by Bruce Cockburn)

What You Will Learn:

- the role of the rain forest in the biosphere;
- ways in which the rain forest is being destroyed;
- the consequences of deforestation;
- some of the products we depend on that come from the rain forest;
- the extent of biological diversity in the rain forest and the potential impact of its loss;
- the effects of rain forest destruction on native groups;
- the variety of perspectives associated with the rain forest issue.

Imagine!

Rain forests are not restricted to the tropics. They also occur in more temperate areas, including Canada, the United States, Chile and Australasia.

Epiphyte growth in a Costa Rican rain forest.

Epiphytes: plants that grow upon other plants, but are not generally parasitic.

Introduction

Tropical rain forests are wet and hot environments, making them the most productive biome in the world. These forests once stretched around the earth's middle, covering 20% of the land. With no defined seasons and constantly high rainfall, conditions in the rain forest are perfect for rapid, year-round plant growth of astonishing diversity. These complex plant communities in turn support a profusion of animal species.

This incredible environment is one of the most threatened on earth. It has been broken into pockets, reduced now to about 7% of the earth's land area, and it continues to disappear at an alarming rate.

Although there are many rain forests throughout the world, this chapter focuses on the largest, the Amazon Basin of South America. This forest covers roughly 7.5 million km^2 (an area about three-quarters the size of Canada) and extends into several countries including Brazil, Peru, Ecuador, Colombia, Bolivia and French Guiana.

The Tropical Rain Forest Ecosystem

Plants and Animals

The tropical rain forest environment is a layered system (see Figure 13.1). Uppermost is a thick *canopy* of interwoven tree tops exposed to sunlight, wind, and rain. Most of the animal life dwells in this layer. A few very tall trees, called *emergents*, rise above the canopy, but despite their great height, none of these trees are deeply rooted. Rather, they have shallow, widespread roots that quickly absorb nutrients from the thin forest floor. Below the canopy is the shadowy *understory*, composed of small trees and shrubs that don't need direct sunlight. Both the canopy and understory contain millions of plants called **epiphytes**—orchids, ferns, bromeliads and cactuses—that grow on other plants. It is these plants that contribute to the forest's impressive diversity of life. The bottom layer is the forest floor, which is dark, humid, quiet and open. Only about 2% to 5% of the sunlight falling on the canopy reaches the forest floor, so few plants can survive there.

The rich and abundant plant growth creates numerous habitats for specialized animals, most of which live above the ground in one or more of the layers. So specialized are many of these animals that they feed on only a few types of food, thus reducing competition with other species. For example, an alga grows in the fur of the slow-moving sloth, giving the fur a greenish tinge. A certain species of caterpillar is adapted to feed only on this alga. The relationships among many of these specialized plants and animals have evolved so that each is dependent on the other. For example, an animal may feed only on the nectar of a single type of flower, and that flower, in turn, relies solely on that animal for pollination. On the forest floor, there are few animals compared to the number that live above it. Ants are among the most common animals at this level, where they feed on the **detritus** which has fallen from above.

Water Cycle and Climate

The rain forest is a major regulator of the earth's water cycle and climate. About half of the high rainfall is taken up by plants within the forest, but the other half becomes runoff that contributes to a very extensive system of rivers and lakes. In continuously high temperatures, the trees transpire constantly, releasing an enormous volume of water which rises, cools and falls as rain again. As the warm water vapour rises into the upper atmosphere, it gives off its heat to the surrounding air, which spreads towards the poles, cools and sinks to earth again. This air circulation has a major effect on the climate of areas thousands of miles to the north and south.

Soils

In contrast to the soils of northern forests, those of tropical rain forests do not have large quantities of nutrients. They are thin and poor, providing little more than a means for anchoring the plants. Most nutrients cycle very quickly and efficiently through the living system rather than building up in the soil. For example, as the

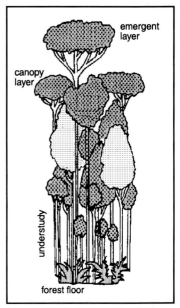

Figure 13.1 - Layered rain forest structure.

Detritus: dead and decayed organic material.

Imagine!

Because of the high transpiration rates, the rain forest is said to make its own rain. About half the rain that falls on the forest has fallen on it before.

Imagine!

About 90% of the nutrients in the rain forest are in the bodies of plants and animals. Only 10% is in the soil.

221

How Many Species on Earth?

The level of **biodiversity** in the rain forest is staggering. In a study by the Smithsonian Institute, scientists found 41,000 species of insects in the tree canopy covering one hectare. The number was so large that it caused them to elevate their prediction about the number of different species on earth from 10 million species to 30 million.

Biodiversity: variety of life in a given area.

Imagine!

A ten hectare tract of a Malaysian rain forest had 750 species of trees. There are only about 700 different species of trees in all of Canada and U.S.

Imagine!

On December 22, 1988, Chico Mendes, an outspoken rubber tapper, was shot for his efforts to defend the rain forests.

remains of plants and animals begin to fall to the forest floor, many are intercepted and recycled by the epiphytes.

Biodiversity

Unlike more temperate and polar areas, where there are many individual members of only a few species, tropical rain forests have many species, but fewer members of each. Scientists conservatively estimate that 50% of the world's species live in the tropical rain forests.

What might account for this incredible richness? This is a topic of much debate among scientists. Is there a great number of species because of the fast recycling of nutrients, or are nutrients recycled quickly because of the variety of species? This question remains unanswered. Yet it is known that some of the major factors contributing to biodiversity are the complex physical structure, age, and stability of the tropical rain forests. These ecosystems are millions of years old, undisturbed by glaciation, fire and cutting, unlike many northern forests. Plants and animals have had a long time to evolve.

The Political, Social and Economic Environment

Since the 1930s, the countries of the Amazon Basin have been struggling to raise their living standards and to become a competitive force in the world economy. To make the necessary technological advances, South American governments have borrowed heavily from banks in other countries. This money has been spent on roads, hydroelectric and industrial projects and military hardware.

In order to pay off their massive debts, these countries are trying to make money fast. Their immediate concern is short-term economic survival, not long-term ecological stability. Survival is also made more difficult by severe social problems around the Amazon. Large numbers of the urban poor are moving out of the cities and settling in the country because of the harsh city street-life. In an effort to survive, they are cutting down trees to obtain fuel and to grow crops.

Underlying these difficulties are very unstable and often corrupt political systems. Money and might influence many government policies. Those who speak out to save the rain forest do so at their peril since many rich landowners will not hesitate to have people killed if they interfere with their source of income.

Activities Causing Deforestation

Cattle Grazing

Perhaps the single biggest cause of deforestation is the conversion of large areas of forest into pasture-land for beef cattle. A forest converted to pasture can produce about 100 kg of meat per hectare during the first year. Within five to ten years, however, the soil fertility is depleted and production decreases to about 10 kg of meat per hectare. Much of the meat produced on this land ends up in the fast food industry and in pet food.

Logging

Although it is a point of debate whether logging is a cause of deforestation or not (see margin), considerable logging does occur in rain forest environments. However, it is practised more intensively in Africa and Asia than it is in the Amazon. Trees from the rain forest, such as mahogany, sandalwood and ebony are exported and used in products for first world markets. The trees are harvested either by selective cutting or clearcutting, both of which can be harmful. Selective cutting involves taking the best and largest trees. But felling a few large trees will inevitably bring down others, because branches are interwound in the canopy and strong vines link trees together. The remaining shallow-rooted trees can then be toppled by the wind. Dragging the huge trees out with heavy equipment causes more damage. Clearcutting is even more disastrous, virtually ruining the land after exposing it to the heavy rains which quickly erode the soil.

Imagine!
Two hectares of rain forest land can support 1200 trees of 200 different species, or one cow.

Deforestation and Logging Debate
Many professional foresters argue that logging in tropical rain forests does not cause deforestation. Since most logging activity involves selective cutting, forests are altered but not permanently removed. In other words, they will regrow. Clearcutting in rain forests is often a result of conversion of the forests to agricultural land, rather than simple tree harvesting.

...According to a recent study commissioned by the International Tropical Timber Organization, not even one tenth of one percent of the remaining tropical forests are being actively managed for sustainable productivity...
...studies in the rapidly deforesting Brazilian state of Acre show that because pastures quickly lose productivity and can carry few cattle, the present per hectare revenue from the collection of wild rubber and Brazil nuts is four times as high as the revenue from cattle ranching...

Robert Repetto
Director of Economic Policies
at the World Resources Institute

223

Imagine!
Major projects such as the Tucutui Dam flood more than 2000 km² of forest and have displaced many people.

Imagine!
The world's largest iron ore deposit is in the Amazonian rain forest.

Slash and Burn Farming

Slash and burn farming, when practised on a small scale, is ecologically sound. A farmer clears a small plot of land and burns the trees and limbs, the ashes of which enrich the soil. Many different plants are sown in this soil, some fast growing, others slower, but soon the plot is covered in vegetation. As the crops mature, they are selectively harvested. After a couple of years, the forest grows back and the farmer moves to a new plot, leaving the original one untended for 10 to 15 years. This style of farming works fine when few people are doing it and when adequate time is given between crops. However, when too many people are involved, and when the same plot is cultivated year after year, the soil is destroyed and the land will bear neither crops nor forest.

Hydroelectric and Mineral Development

Brazil imports considerable amounts of fuel, but in order to support its struggling industrial sector, it is trying to establish its own energy source. By 1990, two major dams had been constructed, flooding about 13,000 km² of rain forest; 31 more hydroelectric projects were planned. The flooding from one of the existing dams, the Balbina, along with a major highway, led indirectly to the decline of Waimiri-Atroari Indians from 3,000 in the early 1970s to about 300 by the mid 1980s.

Interest is also growing in the area's mineral deposits of iron, gold and oil. For example, significant gold deposits were discovered around 1980 near Serra Pelada. By 1983, approximately 40,000 workers had invaded the area in search of the precious mineral. Although the intensity of work has since declined in that area, it has shifted to others. The processing of gold has led to considerable chemical pollution in the Amazon. The largest and richest iron ore deposit in the world is also under development in the Carajas mountains. This project is contributing to the deforestation of 1,500 km² per year.

Rate and Effects of Deforestation

No one is certain how fast the rain forests are disappearing. The United Nations estimated in the late 1980s that the tropical rain forest was being lost at a rate of 0.7% annually. If this rate continued, most forests would be gone or seriously disturbed by the year 2135. The World Resources Institute estimated that one third of the earth's original tropical rain forest has been destroyed and the rate of destruction continues at about seven million hectares per year, or 13 hectares per minute. More recently, in 1991, National Geographic reported the loss to be 14 million hectares per year (an area one-third the size of all Newfoundland and Labrador). Despite the lack of precise agreement on the rate of loss, the conclusion is the same: massive areas of rain forest will be permanently lost in your lifetime.

Rain forest depletion accounts for the greatest threat to global biodiversity. One estimate suggests that we will lose 20% of the earth's species by the year 2000.

With deforestation, the water cycle is also disrupted because the forest no longer traps water. The water instead flows freely across open land, picking up fine particles that will enter the rivers and cause silting. Because the forest no longer retains any water, flooding and erosion occur with the fast runoff. Over a short time, the weak soil loses its nutrients and is unable to support life. The heavy rainfall, partly created by transpiration from the forest, now decreases and the area can become a dry, hot desert, like the Sahara.

With less water vapour rising into the air and condensing, there is a decrease in the heat given off to the surrounding air. This in turn causes a reduction in the air flow that cycles towards the poles and then back towards the tropics (see Figure 13.2). With this change in the air flow cycle, the climate of areas well north and south of the tropics is expected to change.

One of the most distressing impacts of tropical rain forest depletion occurs for the people who make it their home. People of the Amazon, like the Yanomani, have sustained themselves for eons without damaging their environment. But they are now threatened, both by the

A strangler fig growing upward on this tree will eventually kill the host plant (top). The white-faced monkey is a popular sight in Manuel Antonia Park, Costa Rica (below).

Imagine!

"Land distribution in Brazil is among the most unequal in the world. Just 4.1% of landowners have 81% of the farmland, while 70% of rural households are landless."

The New Internationalist, May 1991

loss of their way of life and by the influx of disease. A simple flu virus brought in by a foreigner can sometimes decimate an entire village, since the native people have not built up natural resistance to the common ailments of our modern societies. The loss of these native peoples is a loss to people everywhere. Within these ancient forest societies, there is a storehouse of knowledge about the tropical environment, any portion of which might lead to the discovery of valuable new foods and medicines.

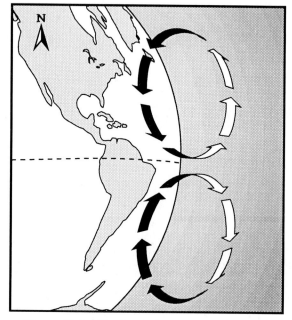

Figure 13.2 - Air circulation over the equator.

Analysis:

1. *Under the following headings, compare the structure of the tropical rain forest with the boreal forest in Newfoundland: a. nutrients   b. biodiversity   c. layering*

2. *Describe how tropical rain forests influence climate.*

3. *Discuss the following statement: "Third world countries are rapidly harvesting their natural resources so that they can pay off their large debts to the industrialized world."*

4. *Outline the major causes of rain forest depletion. Compare these problems with those confronting forest managers in Newfoundland or Canada.*

5. *If Canada and the United States have already harvested most of their original forests, should countries like Brazil be allowed to do the same? Support your answer with information from the text.*

226

"...the poor...put pressures on the forest but you cannot blame them. It is an unstoppable phenomenon, connected with poverty. They may ask, "Who are you conserving the forest for? We have to survive." This needs big programs of poverty alleviation."

Dr. Y. Sudhakara Rao,
Forestry Official,
Food and Agriculture
Organization, 1991

"Efforts to stop the destruction run into moral as well as practical obstacles. How can developed nations demand onerous debt payments and ask the debtors to preserve the forests? How can countries worry about biodiversity when their people are concerned with feeding themselves?..."

Eugene Linden
Time Magazine, 1989

Perspectives on the Rain Forest

"Viewed from Brasilia, some of the demands of the western greens [environmental groups] seem ludicrously unrealistic. Amazonia ...covers more than half the country's land area. It contains the world's richest deposit of iron ore, and other mineral resources that have barely begun to be explored. Some of it could undoubtedly be turned into productive farm land. Should it be left unexploited?"

Author unknown
The Economist, 1989

"Not so long ago...some Kampla [an Amazonian tribe] felled 12 majestic, century old mahogany trees to pay for just one shotgun and ammunition— now a necessity because overhunting by white settlers has reduced the availability of game beyond the point where a bow and arrow can produce a full stomach."

Alex Shankland
The New Internationalist ,1991

In The News...

The following article extracts from newspapers and magazines represent a range of viewpoints on the tropical rain forest issue—some looking at causes, others at solutions. While the political factors are being analyzed, and hopefully resolved, we can also be doing our part as consumers to help stop the destruction of these valuable habitats.

Extract from

Deforestation in the Tropics

by Robert Repetto. *Scientific American*; Vol. 262, No.4, April 1990.

"In the Philippines, Brazil and elsewhere, recognized rights of occupancy or possession are awarded on the basis of the area of land cleared. Such provisions often become a mechanism for privatizing land from the public forest estate. Those who obtain ownership soon sell out to larger capitalists, who consolidate the land to establish private ranches and accumulate speculative holdings.

In many cases such activities would be uneconomic without heavy government subsidies. In the Brazilian Amazon, road-building projects financed by the federal government and multinational development banks have fuelled land speculation. More than 600 cattle ranches, averaging more than 20,000 hectares each, have been supported by subsidized long-term loans, tax credits covering most of the investment costs, tax holidays and write-offs. The ranches proved to be uneconomic, typically losing more than half of their invested capital within 15 years.

Indeed, surveys showed that meat output averaged only 9 percent of what was projected and that many ranches were reorganized and resold repeatedly, having served only as tax shelters. (In that respect, they were unquestionably productive, generating returns of up to 250 percent of their owners' actual equity input.) Although the Brazilian government has suspended incentives for new cattle ranches in Amazonian forests, supports continue for existing cattle ranches, covering 12 million hectares, that have already cost the treasury more than $2.5 billion in lost revenue."

Analysis:

6. *Describe the mechanism that the Brazilian government uses to privatize land. Is this a good method? How might this method be abused?*

7. *How might government use incentives to encourage preservation of the rain forest rather than destruction of it?*

8. *Research the regulations regarding lease of crown land in Newfoundland and Labrador.*

Gathering Strength

by David Ransom. *The New Internationalist*, May 1991. Reprinted with permission.

"...But what can people outside the country do? "The most important thing you can do...is to be willing to learn more and understand the problem, because it is only through understanding that you will find out what has to be done."

When I set out on this trip I was anxious to discover a model of 'sustainable' development, a way of protecting the forest while improving the lives of the people, that we might all (smugly?) subscribe to. What I actually found out is that no such model yet exists. The best that experts can provide for us is some kind of 'goal', a framework of values yet to be filled with knowledge and experience.

Consider this. Everywhere in the Amazon the destruction of the forest has followed the building of roads. The BR 364 has ripped through Rondônia and there are plans to drive it on through Acre to Peru and the Pacific, opening the veins of the Amazon to the thirst of the Japanese for hardwoods and raw materials. José Lutzemburger, the celebrated environmentalist who is now Brazil's Minister for the Environment, has gone on record to say that the road will not go through. Surely, I thought, any self-respecting environmentalist must welcome this.

But no one I spoke to in Acre either believes or agrees with it. "There's nothing wrong with roads, you know," Macedo said to me. "It's who and what they are *used* for that matters."

I think he's right. You can't condemn the people of Acre to live forever in the nineteenth century. If you do they will simply leave the forest and there will be no-one to protect it. Banning roads in the Amazon doesn't quench the rich world's thirst for drugs and hardwoods, bring about agrarian reform or tackle the gross injustice of Brazilian society.

...The point is that simple, sure-fire prohibitions in response to complex human situations can lead us to false conclusions that may actually make things worse.

Many of the people I met were very anxious indeed to develop markets for 'sustainable' products that don't damage the forest - and, above all, that can provide a decent income. Brazil nuts are one of the most important of these. So yes, go out and buy as many as you can from the co-operatives that produce them....

...The point is that in the process you have taken an active interest in what is going on—become 'engaged'. And that is what *everyone* I met in the Amazon, without exception, wants you to do. You might go on from there to support an immediate ban on the trade in tropical hardwoods, which is urgently needed, or seek out beautiful Kampa handicrafts when and if they reach the market. And you might go on from *that* to wonder why your own government may be supporting the *status quo* in Brazil and demanding the blood of the Amazon in payment of its ill-gotten debts."

Analysis:

9. How are road building and rain forest destruction related?

10. Should large stands of rain forests be left untouched? Explain your answer.

11. Why do outsiders need to be careful in proposing solutions to other people's problems?

Last Thoughts

Imagine!

Wild yams that grow in Mexico and Guatemala were used to develop the contraceptive pill. They are still used in India and China to produce oral contraceptives.

Rain forest depletion is a global problem, not just a local one. It is easy for us in the relatively rich, developed countries to ask, "Why do they do it? Why don't they just stop?" But the problem is a complex one, influenced by many economic, political, social and environmental factors. The responsibility for fixing the problem does not rest solely with the people of the Amazon or any other rain forest. We must take responsibility as well. How can we expect poor countries to bear all the costs when we provide the markets for many of the products that contribute to rain forest destruction?

There are some signs of improvement in the Amazon. Brazilians are fighting back against political corruption. In September 1992, the Brazilian President, Fernando Collor de Mello was thrown out of government after corruption charges were laid against him. In the world of finance, the World Bank has established a policy that it will fund projects in tropical countries only if the countries applying for loans show a commitment to conservation. If natural areas are lost as a result of a project, areas of equal size are to be protected. The World Bank's perspective on this is that good economic management can only occur along with good environmental management. It is this perspective that provides the basis for sustainable development.

Students in Action

1. Become more familiar with the issues surrounding tropical rain forest depletion. Donate money or time to an organization that is working to help the people and resources of the rain forest.
2. Visit a zoo or botanical garden on your next holiday and find out more about the plants and animals that live in the tropical rain forest.
3. Avoid buying new furniture and other wood products that are made from tropical hardwoods like teak and mahogany. Instead, consider buying and restoring used furniture so that the wood already harvested will not be wasted.

CHAPTER 14
Acid Precipitation

This was no local problem. Rather, the threat mocked provincial and national boundaries and struck to the heart of global issues of sustainable resource use and economic development.

(*The State of Canada's Environment*, 1991)

Introduction

Remote wilderness lakes nearly devoid of life shocked many Canadians during the 1970s into realizing that acid rain was a fact of our lives. The haunting call of the loon, a symbol of the Canadian wild, was silenced in lakes that were previously rich with fish, the loon's main food. Although the effects of acid precipitation were first noticed in freshwater environments, these effects extend into many other ecosystems and aspects of our lives. Science is still struggling to understand the precise influence of acid precipitation on forestry, agriculture and human health.

It was in industrialized England during the 1870s that acid precipitation was first noticed. Robert Smith, a British chemist, found that smoke and fumes caused chemical changes in precipitation, not only near the source of pollution, but considerable distances away. He called the precipitation 'acid rain.' Concern about its widespread effects began to grow in the 1950s and 1960s when people began to notice a decrease in the number of fish in many lakes in eastern Canada, northeastern U.S. and Scandinavia. When tested, these lakes proved to have a higher than normal acidity.

We now recognize *acid rain* as precipitation and particles that have been made acidic by air pollution. The term *acid precipitation* is preferred to acid rain because precipitation can also occur as snow, hail, and freezing rain.

What You Will Learn:

- the difference between 'acid precipitation' and 'acid deposition';
- an understanding of the pH scale as an indicator for measuring acidity or alkalinity of a solution;
- the components of acid precipitation and their sources;
- the global impacts of acid precipitation;
- problems caused by acid precipitation on soil, vegetation, freshwater environments, "built" environments, and human health;
- the political controversy surrounding this issue;
- the intent of international agreements aimed at reducing emissions that contribute to acid precipitation;
- ways in which the acid precipitation issue is being addressed.

What is Acid Precipitation?

Normal rain water is slightly acidic, with a pH of about 5.6. However, scientists have found that precipitation over much of Europe and eastern North America has a pH of about 4.5 to 10 times more acidic than normal. The chemicals found in the precipitation are sulphuric and nitric acids, which are byproducts of chemical reactions that occur in the earth's atmosphere.

Here is what happens. Sulphur dioxide (SO_2) and oxides of nitrogen (NO_x) are chemicals produced by human activity that lead to acid precipitation. Over 90%

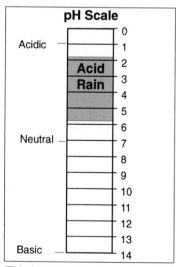

This is a logarithmic scale, which means that a one-unit change up or down the scale represents a change by a factor of 10 in the acidity or basicity of a substance.

of these chemicals come from the smelting of sulphur-bearing ores and the burning of fossil fuels. A major source of sulphur dioxide is coal- or oil-burning electrical generating plants. The internal combustion engine of our cars and trucks is the major source of nitrogen oxides.

Once in the atmosphere, the sulphur dioxide first takes on extra oxygen to become sulphur trioxide (SO_3), which in turn reacts with ordinary water vapour in the air to produce sulphuric acid (H_2SO_4). The chemical equations for these reactions are:

$$2SO_2 + O_2 \rightarrow 2SO_3$$
$$SO_3 + H_2O \rightarrow H_2SO_4$$

The nitrogen oxides follow a similar path to become nitric acid. For example, nitric oxide changes like this:

$$4NO + 3O_2 + 2H_2O \rightarrow 4HNO_3$$

These acids in the atmosphere collect in clouds and eventually fall to earth as some kind of precipitation.

Those that do not return as precipitation will often settle out as dust in a process called *dry deposition*. But when they reach a wet or moist environment such as a lake or soil, the same chemical reaction will occur. Whenever we discuss acid rain or acid precipitation, remember that dry deposition is also part of the problem. In fact, the term acid deposition is used to cover both wet precipitation, dry fallout and acid fog.

Effects of Acid Precipitation

Freshwater Environments

It was the lack of fish in lakes throughout eastern Canada, the northeastern U.S. and Scandinavia that set off the alarm about acid rain. This was because the effects of acid rain are most easily detected in freshwater environments.

Aquatic organisms have adapted to a more or less constant pH level. When the pH level changes, and remains changed for a number of years, many organisms

Acid Fog

Of all the types of acid deposition, acid fog is the most acidic. Fog is a cloud that occurs at ground level. In clouds, water vapour condenses on dust particles that contain sulphur dioxide and nitrogen oxides. Through the chemical reactions shown above, this combination forms tiny, concentrated droplets of acid—an acidic cloud. The cloud has a higher acidity than the rain or snow which falls later, because as more water vapour condenses in the cloud, the acidic droplets become diluted. When these clouds meet the land as fog, exposed plants are especially vulnerable, the greatest impact being in high elevation and coastal areas.

Imagine!

A 1981 rainstorm in Baltimore had a pH of 2.7, about the same acidity as vinegar and 1,000 times more acidic than normal rain water. A 1986 fog in southern California had a pH of 1.7, about the same acidity as the hydrochloric acid used in toilet bowl cleaners and 10,000 times more acidic than normal rain water!

In water with near-normal pH, lake trout are plump (above). But after the water becomes acidic, food is so scarce that the long-lived fish starve until they almost resemble eels (below).

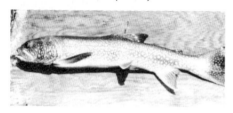

___

Imagine!

More than 80% of all Canadians live in areas where acidic deposits exceed acceptable levels.

___

living in the water will die. Certain kinds of zooplankton, which form the basis for the food chains in freshwater systems, die with slight increases of acidity. Shellfish and finfish, like trout and salmon, are also very sensitive. As acidity increases, more species of zooplankton, insects, crustaceans, amphibians and other fish are affected. In lakes with a pH of about 4, the only aquatic organisms that can live are certain types of algae. These lakes look deceptively pure and blue, but below the surface there is no life.

Sudden changes in acidity, even if they do not continue for long periods, are equally harmful. *Acid shock* can occur with spring runoff after high levels of acid have accumulated in the snow all winter. When it melts, the acidity of streams, rivers, ponds and lakes increases substantially. This can cause a sudden die-off of many organisms, including fish.

Because of the interdependence of organisms living near freshwater environments, song birds, ospreys, eagles, ducks, beavers, otters, and many other types of wildlife are eventually affected by the acid precipitation.

Another problem in aquatic ecosystems involves the effect of acid rain on aluminum, a common mineral found in rock and soil. Aluminum is toxic to fish, but normally it does not dissolve in water and is therefore harmless. However, when the rocks and soil are attacked by acid precipitation, the aluminum dissolves into solution and enters the water. This process, known as *metal mobilization*, can also occur with other elements, such as lead, copper and mercury.

Analysis:

1. *When was acid precipitation first identified as a potential problem?*
2. *What are the two major chemicals causing acid precipitation? How do these chemicals produce acids in the atmosphere?*
3. *Describe three types of acid deposition.*
4. *If higher acidity in an aquatic ecosystem directly affects only one or two organisms, why is the whole ecosystem threatened?*
5. *What is acid shock?*
6. *How might mobilization of metals be a threat to human health?*

Forests

The effects of acid rain on forests were first detected in central Europe. West Germans found extensive areas of damaged forest in the early 1970s. In 1980, 60% of the trees in an area between East Germany and Czechoslovakia were healthy. Two years later, 98% were dead or dying. Similar observations have been made in the forests of the northeastern U.S. In parts of Ontario and Quebec, the growth rates of spruce, fir, and pine have decreased by 50%.

Some scientists argue that a lot of forest death is caused by other factors such as insect damage, but it is generally agreed that acid rain at least contributes to the environmental stress that trees and plants must endure. As plant communities in forests are altered by acid precipitation, wildlife populations will change as well.

Human Environments

Buildings all around the world, some of which have stood for hundreds of years, are dissolving before our eyes because of acid precipitation. Limestone, sandstone and marble—common building materials in cities everywhere—are especially vulnerable. The Acropolis, a symbol of western civilization in Greece, is crumbling. Statues and monuments that embody some of the world's finest art are becoming formless as arms, eyebrows, or locks of hair dissolve under the ruining rain. Others made from bronze are pitted and stained. Bridges require constant testing to ensure that the lime used in the concrete is not dissolving, while pipes carrying water to our homes are becoming corroded by acids in the water that flows through them.

Acid rain does not seem to have an immediate impact on human health. However, the inhalation of dry fallout containing sulphur and nitrogen may lead to respiratory problems. Increased leaching of toxic elements like aluminum, copper, lead and mercury from the soils into our water systems is also potentially harmful. Many of the links between acid precipitation and human health are poorly understood at this point, but research in these areas is continuing.

Effect of Acid Rain on Plants

Acid precipitation falling directly on plants will leach out many of the nutrients from the leaves. In the soil, it will wash out essential plant nutrients, such as potassium and calcium, making them unavailable to the plants. Acid precipitation also breaks down the protective covering on the leaves of many plants. Many scientists believe that this makes them more vulnerable to attacks from insects, fungi, and bacteria. Dissolved aluminum, discussed earlier, is also a serious problem for many plants.

Imagine!

Acid precipitation is both an environmental and economic problem. During the early 1980s, the German lumber industry estimated an $800 million per year loss because of acid precipitation. The value of crops at risk in eastern Canada was estimated at $3 billion.

Imagine!

Most of Newfoundland and a considerable portion of Labrador are highly sensitive to the effects of acid precipitation because the soils are already acidic and they lack the buffering capability of many other areas.

Imagine!
84% of the most productive agricultural lands in eastern Canada annually receive more than the acceptable levels of acid deposition.

Liming Lakes

Either before a body of water becomes too acidic or after this happens, lime can be added to neutralize the effect of acid precipitation. When done before, the addition of lime can reduce the water's sensitivity to acid precipitation.

This process works because the pH of lime measures near the basic end of the scale. Therefore, it can buffer acidity in a lake or pond in the same way that it does in a garden or farmer's field. Natural limestone bedrock in certain areas can have the same effect.

The major problem with this technique is the scale with which it must be applied, and hence the associated cost. Imagine the difficulties involved with getting tonnes of lime into remote wilderness lakes.

Sweden has used this technique extensively, but in Canada it has been rejected because of the technical difficulties and because it does not provide a permanent solution to the problem.

Different Impacts on Different Areas

Not all natural systems react in the same way to acid precipitation. Some lakes, forests and soils seem to have at least some natural resilience to acidification. These are usually areas with high concentrations of limestone (calcium carbonate - $CaCO_3$) in the bedrock. Dissolved limestone can provide a neutralizing base, or *buffer*, to the effects of acid precipitation. However, the buffering ability of a system can suddenly be exhausted. When that happens, even a small increase in acidity can cause a sudden, large drop in pH, which can kill a system very quickly. This is what may have happened with the European forest between 1980 and 1982.

International Agreements to Reduce Acid Precipitation

Anywhere in the world where there is significant industrial pollution, there are problems with acid precipitation. Because this pollution is rarely contained within the country where it originates, it is known as *transboundary* pollution. Although the effects of acid precipitation have been painfully felt in Scandinavia, the source of most of the pollution is in Poland, former East Germany and Czechoslovakia. Some of Canada's acid precipitation originates here, but the vast majority of it is from the U.S.

In 1970, the World Meteorological Organization and other concerned groups of scientists began to search for solutions, and to make people more aware of the problems associated with transboundary pollution. In 1982, this search for solutions led to an international conference in Stockholm, Sweden. The Stockholm meeting was followed by others in 1984 and 1986. International negotiations on acid rain are always difficult, often because the countries creating most of the pollution are very concerned about the high costs of controlling it. Canada, which was one of the more vocal countries arguing for better pollution controls, found itself allied with Scandinavia and West Germany in this cause. The conferences resulted in the signing of the Thirty Percent Protocol—a statement of intent signed by 21 countries to reduce sulphur emissions by 30% below 1980 levels. This reduction was to be accomplished by 1993. The United Kingdom, the United States and Poland, three of the most significant polluters, refused to sign the agreement. France, Canada, Norway, Sweden, Denmark and West Germany agreed to reduce their emissions by as much as 40% to 50% (see Figure 14.1).

While Canada has been involved in negotiations with many countries, its main concern has been with the U.S. Emissions in both countries have to be reduced to have any significant effect on Canadian lands and water systems. Canada eventually reached an agreement with the U.S. The strategy involved reducing our own emissions first. This allowed Canada to argue its case with the United States more effectively in trying to get the U.S. to reduce its emissions as well.

Finally, the Air Quality Accord was signed in 1991 (see margin). Scientists expect that the reductions in emissions required by the Accord will stop further acid rain damage in Canada, with the exception of very sensitive areas in Atlantic Canada.

The Air Quality Accord requires:

- the U.S. to cap SO_2 emissions at 13.3 million tonnes;
- Canada to cap its total SO_2 emissions at 3.2 million tonnes by the year 2000;
- both countries to reduce NO_x emissions from factories and power plants according to set schedules;
- both countries to establish tighter emission standards for the motor vehicle industry;
- both countries to monitor closely their emissions of SO_2 and NO_x;
- both countries to report publicly on their progress to reduce SO_2 and NO_x emissions;
- both countries to take necessary actions to protect the air quality around protected natural areas such as national and provincial parks and wilderness areas.

Imagine!

Nearly one quarter of Sweden's 90,000 lakes are acidified. Four thousand are so acidified that fish do not survive in them. In southern Norway four out of five lakes and streams are dead or dying.

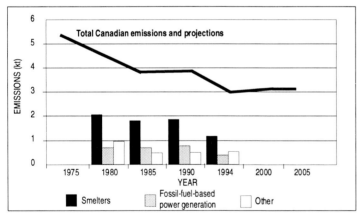

Figure 14.1 - Actual and projected emissions of sulphur dioxide in eastern Canada, 1975-2005.

Analysis:

7. Discuss the statement: *"Air pollutants respect no borders."*

8. Find out what alternate technologies are available to help reduce emissions that lead to acid precipitation.

9. Why would the U.S. and U.K. be reluctant to sign agreements to reduce acid precipitation?

"There's no consensus in the country (U.S.) about acid rain... We (have not arrived) at the point where the country is sufficiently convinced that we have a real problem here that we have to address..."

William Ruckelhaus
Environmental Protection
Agency, 1984

"The standard scientific view of acid rain's effects may simply be wrong...In a study of acid rain, my colleagues and I calculated that it could eventually cost Americans 100 billion dollars in today's dollars to achieve a major reduction in sulphur dioxide emissions. Before committing to any program of this magnitude, we should want to be more certain that acid rain is in fact a major threat to the country's environment."

William Brown
Fortune Magazine, May 28, 1984

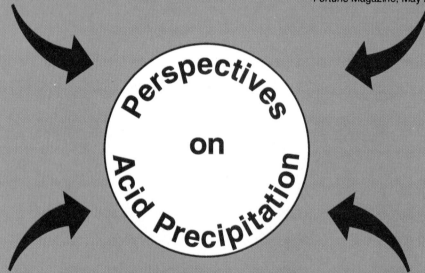

"In balancing the pros and cons of instituting controls, it seems to me that the issue boils down to one of values more than to information that science can provide. Those who believe that pride in practising good stewardship for the environment is worth the monetary cost can demonstrate the reality of adverse environmental effects. Those who insist that more needs to be known before controls are instituted do so because they are willing to take risks with the future environment in order to avoid near-term economic costs."

Arthur H. Johnson
soil scientist

"We have enough information to act...It's not a matter of science any longer, it's a matter of political will. We have reached the point where a decision to only do more research is, in fact, a decision to do nothing."

Canadian Environment Minister
John Roberts, 1983

In the News...

These following article extracts explore two orientations related to science and the problem of acid precipitation. The first highlights some of the inadequacies of science in identifying the problem; the second looks at an example of a scientific approach to dealing with the damage caused by acid precipitation.

Extract from

Maybe Acid Rain Isn't the Villain

by William Brown. *Fortune* Magazine, May 28, 1984

"The possibility that our acidified lakes got that way naturally is hard for people to accept precisely because of this logical difficulty. If some of the fish are now in a more hostile environment than they were in earlier decades, then we must look to something new in the environment. Sulphur dioxide from heavy industry seems to be just the kind of suspect that makes sense. In fact, however, these emissions are not the only change in the forest environment. Another new feature is Smokey the Bear. Or, less metaphorically, the huge success of the United States in preventing forest fires during the past half-century or so.

Forest fires can have a tremendous impact on the acidity of adjacent lakes. The fires can totally destroy the acid-producing humus [a natural filter in the watershed], replacing it with a layer of alkaline ash. When that happens, a naturally acidified lake within the burned area may become neutralized and temporarily—meaning for several decades—more hospitable to fish. Eventually, of course, the forest would be expected to regrow, the alkaline ash left by the fire would be used up, the acidic humus layer would be regenerated, and the fish would be in trouble again.

The possibility that fire prevention accounts for a major portion of the lake's acidity still has to be viewed as just that–a possibility. It's another of the many fascinating hypotheses that are still too new to have been tested properly by field researchers undertaking controlled experiments. Meanwhile, all we know for sure is that the fires are far less prevalent than they once were."

Analysis:

10. *Evaluate William Brown's hypothesis for an increase in acidification of freshwater systems. Do you think his argument has merit? Why or why not?*

11. *Given the scientific controversy of this issue in the 1980s, how would you, as a politician at that time, have responded to potential spending increases to combat acid rain?*

239

Aurora trout...to be reintroduced to former lake habitat

by Craig McInnes. *The Globe and Mail*, May 22, 1990. Reprinted with permission.

"A variety of trout driven to the brink of extinction by acid rain is being reintroduced into its former habitat in Northern Ontario this week.

The Aurora trout, named for its brilliant colours, existed only in three tiny lakes in Lady Evelyn-Smoothwater Park. The park is about 120 kilometres northeast of the nickel smelters in Sudbury, a major source of acid pollution.

Those lakes were heavily acidified by the early 1960s.

"They did a lot of test netting and the lakes were entirely devoid of fish at that time," John Gunn, a biologist with the Ministry of Natural Resources, said in a telephone interview. "It reached extinction in its native waters."

As the lakes deteriorated, a biologist with the Ministry of Natural Resources started raising the Aurora trout in captivity to save the fish...from extinction.

Now biologists with the ministry believe that enough progress has been made in reducing acid rain for the trout to live again in the lakes.

"Last fall we decided we have general good news from the Ministry of Environment that there is a region-wide improvement in water quality associated with reductions in emissions from the smelters," Mr. Gunn said.

The water is still too acidic for the fish to survive without help, so ministry officials have put lime into Whirligig Lake to reduce the acidity.

"It is nothing more sophisticated than adding a Tums to the water. It is an antacid." Mr. Gunn said.

Officials hope that reductions in emissions required over the next few years under provincial regulations will make the lakes naturally more hospitable to the fish.

...But their survival is not certain. Although the lake water can be buffered in the short run with the addition of more lime, in the long term survival depends on rain becoming less acidic.

Even then, it could take decades for the lakes to recover fully.

And it is not clear that the trout, which has been raised in captivity for several generations, can still survive in the wild. Several earlier attempts to stock other lakes with the fish have failed, Mr. Gunn said.

One thing the trout will not have to contend with is anglers. Whirligig Lake has been closed to fishing for more than 20 years."

Analysis:

12. *How does the re-introduction of one species of fish demonstrate hope for the future with respect to air and water pollutants?*

13. *Make a list of the potential gains and losses associated with: a. maintaining the air emission standards of years gone by; b. reducing emissions and restoring the natural environment close to the way it was.*

Last Thoughts

It took many years for scientists to work out the causes and effects of acid precipitation. There is still not full agreement on whether or not it is a serious problem, nor whether the solutions are worth the costs. This raises important questions in environmental science. How much do we need to know before we act? If we wait until we are absolutely certain about the causes and effects of problems, will it then be too late to fix them? Our scientific and political leaders are constantly faced with these questions.

In the case of acid precipitation, there is fairly broad acceptance now that it is a problem, and there is reasonable understanding of its causes and effects. We also know that most of the technology required to reduce emissions causing acid precipitation is available, although costly. This puts the responsibility for solving the problem more into the field of economics and politics rather than in science. However, science is still important in determining the effects of acid rain and in monitoring industries as they try to make improvements.

Although sulphur dioxide emissions are dropping as legislation becomes tougher and new technology is adopted, the volume of nitrogen oxide in the atmosphere remains relatively constant. The root of this problem lies in our romance with the automobile. In attacking the problem, many large cities are beginning to set stiff regulations and restrictions on its use. The hope is that we can reduce our reliance on the automobile and adopt modes of transportation that will be better for us and the environment.

L.A.'s Solution

Los Angeles, a city once synonymous with air pollution, has recently proposed a 20-year "smog war" that will change the southern California lifestyle forever.

It calls for raising parking fees, eliminating drive-through banks and restaurants, banning barbecues, encouraging residents to work at home or to live near their jobs and phasing in vehicles powered by electricity, methane and other clean fuels.

Equinox magazine,
July/August, 1989

Imagine!

In 1983, the Canada-U.S. dispute over acid rain became so heated that the U.S. Department of Justice ruled that a National Film Board (Canadian) film on acid rain be labelled political propaganda. U.S. groups that viewed the film were reported to the authorities.

Students in Action

Think of as many different ways as possible to minimize the use of cars and pollution by cars in our daily lives. Write these ideas in your journal; send a copy to your local newspaper; give a copy to your parents...and try to live by your suggestions!

CHAPTER 15

Ozone Depletion

What does it mean to redefine one's relationship to the skies? What will it do to our children's outlook on life if we have to teach them to be afraid to look up?

(Former U.S. Senator Al Gore)

Introduction

Before the 1990s, weather forecasts rarely gave an index of **ultraviolet radiation (UV)** warning people about excessive exposure to sunshine. Forecasters of the 1990s regularly provide this information because of changes occurring with the ozone layer high in the stratosphere.

Ozone is a pale blue, poisonous gas that forms when oxygen molecules (O_2) are split to form single oxygen atoms. These atoms are then attracted to, and combined with, oxygen molecules (O_2) to form ozone. The energy to split the O_2 molecule can come from sources such as lightning, certain portions of sunlight or a machine like a photocopier.

In the early 1900s, scientists realized that there was a thin layer of ozone about 15 to 50 km above the earth in a part of the atmosphere called the stratosphere. It shields the earth from lethal levels of ultraviolet (UV) radiation that could kill all life on land, and much of that in the sea if unchecked. In the 1980s, scientists were shocked to discover an immense 'hole' in the ozone layer above Antarctica. At first it was unknown whether or not this was a natural condition or if it was related to human activities on the planet. Since that time we have learned that chemicals called *chlorofluorocarbons* (CFCs) are contributing to a gradual thinning of the ozone layer in many locations above the earth.

High in the atmosphere, ozone benefits life on earth, but at ground level it is quite harmful in high concentrations. This chapter focuses on the problem with ozone depletion in the stratosphere and briefly describes some of the problems with ground-level ozone.

Like the problem of acid rain, there is considerable controversy surrounding our understanding of ozone, its depletion in the atmosphere and the potential consequences of that depletion.

Think About It:
Try to imagine the size of the full stratosphere surrounding the earth from 15 to 50 km out—it is an immense volume of space. Now imagine determining differences in the concentrations of one thinly distributed gas–ozone–in the various parts of the stratosphere—not an easy task!

What You Will Learn:

- the nature of ozone and how it is formed;
- the role of the ozone layer in protecting life on earth from ultraviolet radiation;
- the chemicals that cause depletion of the ozone layer;
- the current products that contain chlorofluorocarbons (CFCs);
- the role of CFCs in the depletion of the ozone layer;
- the chemistry involved in the ozone layer's breakdown;
- geographic differences in the effects of ozone depletion;
- an understanding of the hole in the ozone layer over Antarctica;
- the effects of ultraviolet radiation on human health, plants and animals;
- examples of actions taken around the world to phase out the production and use of ozone-depleting products;
- the problems experienced by developing countries in phasing out ozone-depleting chemicals;
- alternatives that could help protect the ozone layer.

Ultraviolet radiation: part of the natural radiation from sunlight. The wavelengths have slightly higher energy and are slightly shorter than visible light. It is absorbed by proteins of living tissues where it can break down some of the chemical bonds of these tissues.

243

Nature and Formation of Ozone

Ground-Level Ozone
About 90% of the earth's ozone is in the stratosphere; the other 10% is at ground level. One of its sources at ground level is associated with combustion, especially from automobile engines. Through complex reactions between sunlight and nitrogen oxides and volatile organic compounds (VOCs) in the automobile exhaust, ozone forms as one of the components of smog. A highly toxic substance, the ozone is very harmful to plant growth. Some ozone is made for industrial purposes as well. It is used for killing harmful bacteria, for bleaching, removing unpleasant odours from foods, sterilizing water and creating other chemicals. However, once the ozone is free in the lower atmosphere, it lingers about, trapping heat near the planet's surface.

Ozone is created mainly by the reaction of oxygen molecules to high energy ultraviolet radiation. The UV radiation splits oxygen molecules, which then recombine as ozone. It would seem, therefore, that in order for the ozone layer to have first formed, there had to have been oxygen (created by plants) in the atmosphere. This plant growth likely occurred in the oceans, below the depths where harmful UV radiation could reach. Then, as more and more oxygen was produced, ozone was created. It is believed that by the time life began to evolve on land, the ozone layer had developed and was filtering out about 99% of the harmful UV radiation.

The splitting and recombining of oxygen atoms in the upper atmosphere forms more than ozone. Some of the lone oxygen atoms combine with ozone molecules to recreate molecules of oxygen. The amount of ozone in the stratosphere at any one time represents a balance between these two reactions; one forming ozone, the other breaking it down again into oxygen.

Although the ozone layer absorbs most of the UV light, some passes through to the lower atmosphere, causing our skins to tan. But exposure to too much UV radiation can cause skin cancer and cataracts. It can also weaken the body's immune system, damage crops, and kill plankton, the basis for life in aquatic food chains.

Not only does the ozone layer shield life on earth from excessive UV radiation, it also has a significant influence on climate. As it absorbs UV radiation, it also heats the stratosphere, which in turn affects global air circulation.

What's the Problem?

Imagine!
More than 2/3 of the plant species tested for their reaction to UV rays have proven to be damaged by it.

Chlorofluorocarbons, which were first developed as coolants for refrigerators and air conditioners, were eventually identified as the likely cause of ozone depletion. Some of the other substances linked to the problem are carbon tetrachloride (CCl_4), formerly a common chemical used in the dry cleaning industry; nitrogen oxides (NO_x), from supersonic jets, auto exhaust and fertilizers; and a family of chemicals known as halons, commonly used in fire extinguishers.

In 1973, two scientists—Rowland and Molina—discovered that CFCs could destroy the ozone layer in the stratosphere. When released from refrigerators or other products, CFCs rise into the upper atmosphere where they are split by sunlight, causing them to release free chlorine atoms. These chlorine atoms then begin to interfere with the balanced oxygen/ozone reactions discussed earlier. The chlorine reacts with the ozone by stripping away one oxygen atom to create chlorine monoxide (ClO) and oxygen (O_2). The unstable chlorine monoxide molecules are easily broken again and again by sunlight, freeing the chlorine atoms to destroy more and more ozone molecules.

These reactions are represented by the following chemical equations:

$$Cl + O_3 \longrightarrow ClO + O_2$$
$$ClO + O_3 \longrightarrow Cl + 2O_2$$

The result of these reactions is:
$$2O_3 \longrightarrow 3O_2$$

It wasn't until 1985 that an atmospheric scientist, Joseph Farman, reported his observation about significant ozone depletion over Antarctica (see Figures 15.1 and 15.2). This shocked the scientists working on the problem because they had not realized that a major loss of ozone had already occurred. It became known as the 'hole' in the ozone layer, but was actually a very low concentration of ozone rather than a hole. The depletion was later confirmed by a number of other scientists, and several theories were proposed to explain its presence. The one that gained most acceptance was suggested by Farman, based on the work done several years earlier by Molina and Rowland. He blamed the hole on chlorine, the main source of which was CFCs.

The following theory was eventually proposed. Winds in the stratosphere carry the ozone-depleting chemicals to the poles, where they become trapped in the swirling winds over the Antarctic called the *polar vortex*.

Imagine!
The problems associated with ozone depletion in the upper atmosphere and ozone accumulation at ground level are quite independent of each other.

Uses of CFCs
- coolants in refrigerators and air conditioners
- propellants in aerosol cans like hair spray and air fresheners
- cleaning agents for circuit boards
- foaming agents in furniture, insulation, carpets and styrofoam containers

---

Imagine!
Variation in solar activity (sunspots) can cause a 10% to 30% change in the UV radiation reaching the earth.

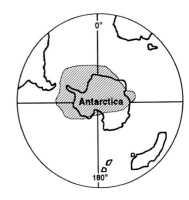

Figure 15.1 - The ozone 'hole' over Antarctica as it appeared in October of 1989.

Imagine!
A sun tan is our skin's way of protecting itself from harmful UV radiation.

We now know that the ozone layer can be as thick as 30 km, but the amount of ozone in that layer is amazingly sparse. At its most concentrated, there are only about 8 molecules of ozone per million molecules of air.

Imagine!
It takes about eight years for CFCs to rise into the upper atmosphere, so it will take many years to see any benefit from reducing our use of CFCs.

Dobson Units
Ozone concentrations are measured in Dobson Units (DU). 100 DU is equal to an imaginary layer of ozone 1mm thick at sea level.

This vortex forms during the sunless winter when the temperatures are extremely cold. During this time the ozone-depleting chemicals collect in stable ice clouds and cause no harm to the ozone layer. Then, when light returns in the polar spring, the chlorine and other ozone-depleting chemicals are split from their previously stable molecules and begin to attack the ozone layer, thus creating the 'hole'. Air samples taken from the area confirm that the concentration of ClO is about 500 times higher than it is at the same elevation at mid-latitudes. The 'hole' lasts about two months, and then fills in as spring advances and ozone-rich air flows in from mid-latitudes.

The seasonal depletion of ozone in the Antarctic is partly a natural process, but our use of ozone-depleting chemicals has accelerated the process.

Ozone depletion occurs over Antarctica and the Arctic, but is less pronounced in the Arctic because it is not cold enough there for the ice clouds to form as often, and air circulation patterns in the region are different. As the ozone-poor air becomes redistributed each spring, it affects regions at mid-latitudes as well.

Although there is general agreement on how these processes occur, they are still theoretical, not fact. We know that ozone depletion occurs at the poles and that it is getting worse. But since our understanding is far from complete, studies are continuing to confirm and refine these ideas.

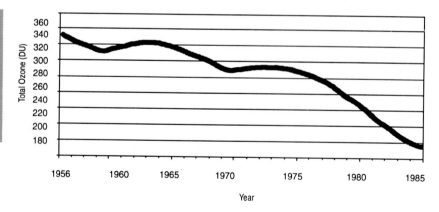

Figure 15.2 - Total ozone values over Antarctica, 1956-85.

Finding Solutions

As the chemical processes causing ozone depletion were being figured out, each new proposal was being doubted and resisted, especially by those with something to lose, such as the companies manufacturing the chemicals and the industries supported by them. Controversy continues to surround the issue of ozone depletion, but evidence against CFCs and other chemicals has built to the point where political action is clearly justified.

Most countries agree that we should act as quickly as possible to reduce our use of chemicals that cause ozone depletion, especially CFCs and halons (see Figure 15.3). However, a huge volume of these chemicals remains contained within refrigerators, air conditioners, and fire extinguishers. We know also that CFCs remain in the atmosphere for a long period of time, and real improvements are a thing of the distant future. It is agreed nonetheless that we must start somewhere.

International Agreements

The first international action to control ozone depletion was taken in 1977 by the United Nations Environment Programme. A World Plan of Action on the Ozone Layer was drawn up, focusing on international co-operation and research. Around the same time, many countries began to restrict the use of CFCs in certain products such as aerosol cans. Given the discovery of the ozone 'hole' in 1985, the Montreal Protocol on Substances that Deplete the Ozone Layer was signed in 1987. It required that by 1998 the 24 countries signing the agreement had to reduce their use of CFCs by 50% and freeze halon use at 1986 levels. Soon after the signing, it was obvious that these controls were inadequate, so more agreements followed. By 1991, 50 countries had agreed to the requirements of the Montreal Protocol. Canada, meanwhile, had decided to go even further and eliminate all CFC production by 1997. Many other countries have since decided to speed up the process of eliminating CFCs as well.

Imagine!

By 1974, Americans were spraying CFCs from aerosol cans into the air at a rate of about 230 million kg per year.

The Problem for Developing Countries

Developing countries find it difficult to phase out CFC use as quickly as the industrialized nations. Because of their warm climates, they are much more dependent on inexpensive refrigeration. The cost of substitutes is high, so poor countries are hit the hardest. Countries like India and China argue that the western nations have caused most of the problem through a much higher level of production and use of ozone-depleting chemicals, so the rich countries should bear the cost of finding and implementing solutions.

In recognition of this concern, the richer countries have agreed to contribute to a $160 million fund to help poorer countries become independent of ozone-depleting chemicals. Also, developing countries are not required to reduce their dependence on CFCs as quickly as richer nations, as long as their use of these chemicals remains relatively low.

Alternatives to Ozone-Depleting Chemicals

It is relatively easy to eliminate or find replacements for CFCs in aerosols, packaging, and foam cushions. However, finding alternatives for coolants and fire extinguishing chemicals is much harder.

Clearly, part of the answer lies in the recovery and reuse of some of these chemicals, rather than their dispersal into the atmosphere. Experts feel that by the year 2000 these actions alone might reduce CFC use by 29%.

Another part of the solution lies in the development of substitutes for CFCs. The most likely candidates are new chemicals called *hydrochlorofluorocarbons* (HCFCs) and *hydrofluorocarbons* (HFCs). The addition of hydrogen causes these chemicals to break down more quickly in the lower atmosphere so they don't pose a threat to the ozone layer. While HFCs contain no chlorine, HCFCs do, although they are not as great a threat to ozone as CFCs. The switch from CFCs to these new chemicals will require significant and costly changes in the machinery and technology used in the refrigeration industry. Ironically, HCFCs and HFCs are powerful greenhouse gases that will increase global warming (see Chapter 16).

Substitutes for halons in fire extinguishing equipment were being developed in 1992, but their effectiveness was still unproven. Although many of the substitutes being developed are far from harmless, they are not as great a threat to ozone as CFCs. Governments and industries have agreed to use these substitutes until more preferable alternatives can be found.

Imagine!

CFCs can survive in the atmosphere for over 75 years.

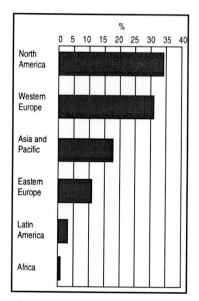

Figure 15.3 - Global consumption of CFCs by region, 1987.

Analysis:

1. *What is ozone, and what is its role in the atmosphere?*

2. *Outline the sources and potential dangers of ground-level ozone.*

3. *Use a diagram to show how CFCs destroy ozone molecules.*

4. *Explain why the ozone hole begins to form towards the end of the antarctic winter, even though CFCs accumulate in the stratosphere throughout the winter.*

5. *Construct a time-line, showing the evolution of the ozone problem from its discovery through to the latest international agreements.*

6. *Find out which appliances in your home use CFCs as coolants. Determine the total volume of these coolants. What happens to the CFCs when these items are discarded?*

"I don't think we're going to be sunburning in five minutes this spring, but the real problem is going to be how we deal with this in the year 2000."

Wayne Evans
Environmental Science Department,
Trent University, 1992

"We already know that ultraviolet light can impair immunity to infectious diseases in animals. We know that there are immunological effects in humans, though we don't yet know their significance."

Margaret Kripke, M.D.
Anderson Cancer Centre, 1992

"Even with an immediate phase-out of all CFCs and related compounds, the world is committed to worsening ozone destruction well into the next century."

Dr. Mostapha Tolba
Executive Director,
United Nations Environment Programme (UNEP), 1991

Perspectives on Ozone Depletion

"It'll take 100 years or so, but we believe the change to the system is reversible."

James Anderson
Harvard University
chemist, 1992

"If you change the amount of ozone or even just change its distribution, you can change the temperature structure of the stratosphere. You're playing there with the whole scheme of how weather is created."

Sherwood Rowland
chemist, University of California
(one of the first to discover the dangers of CFCs), 1992

"The [Montreal] Pact shows a global recognition of the problem but does not even come close to providing the necessary level of protection...There are numerous opportunities for effective industry, government and consumer action. For starters, there are some products which should be simply banned."

Friends of the Earth, 1988

In the News...

The next three pages contain excerpts from magazines that focus on different aspects of the ozone issue. You will read about the realities of living in southern Chile, on the edge of the ozone hole; you will learn about some of the international efforts to deal with the problems; and you will realize the power of individuals in bringing about changes relating to the future of the planet.

Extract from

What the Canadian government should do now

Reprinted courtesy of *Alternatives*, Canada's environmental quarterly since 1971.

At the time of the signing of the Montreal protocol on protection of the ozone layer, federal environment minister Tom McMillan gave a strong commitment to implementing controls in Canada.

Friends of the Earth recommends immediate action on the following specific steps:

- **Food packaging**. The use of styrofoam food packaging and containers should be banned by regulation.

- **Refrigerators and air conditioners**. Regulations should be promulgated requiring all cooling and refrigeration systems to be drained before disposal and the coolant collected and destroyed. No new institutional systems should be built using CFCs and existing systems must install rupture recovery systems. Air conditioning and refrigeration servicing operations should be stringently regulated and inspected.

- **Rigid and flexible foams**. Alternative materials should be promoted. Foam blowing operations should be regulated under the Clean Air Act by setting strict emission limits for CFCs.

- **Aerosols**. Regulation SOR/81-365 under the Environmental Contaminants Act banning the use of CFC propellants in some aerosols should be extended to all aerosols except for medicinal uses.

- **Solvents**. Strict emission limits under the Clean Air Act should be imposed on operations which use CFC solvents. Recapture and reuse systems should be mandatory.

- **Restricting imports**. Declining quotas on the import of products made with and containing CFCs should be developed and enforced. Canada's indirect production of CFCs as a result of imports should be publicly reviewed.

- **Product labelling**. A warning label should be developed and affixed to all products made with or containing CFCs and halons.

- **Living within a budget**. Canada should adopt a target of an 85 percent cut in the production of ozone-depleting substances within five years.

Analysis:

7. Note the date of this article. Research the status of some of these recommendations today. Have they been carried out? Can they feasibly be carried out?

8. What kinds of products could you refuse to buy that would reduce CFCs in the environment? How could you attempt to affect other people's purchasing choices?

Extract from

The Canadian who led the battle to save the ozone layer

by Ian Allaby. *Canadian Geographic*, April/May 1990. Reprinted with permission.

"It was a deeply satisfying day for Byron (Barney) Boville when delegates from 33 nations gathered in Montreal in 1987 to approve phasing out production of chlorofluorocarbons (CFCs).

The delegates had come to sign the Montreal Protocol. That an accord was reached at all is a testimonial to Boville's personal persistence and dynamism. He helped organize the United Nation's first ozone conference in 1980, which led ultimately to the 1987 protocol.

Boville's first important contribution to meteorology was a series of papers describing how weather fronts interact over great distances. Meanwhile, a tour of Canadian arctic weather stations rekindled Boville's curiosity about the upper atmosphere. Evidence was accumulating that there were powerful winds, vortexes (whirling masses of air), sudden warmings, and inexplicable fluctuations in ozone amounts in the stratosphere above the North Pole. Not all this tumult could be ascribed to solar influence. Even in the sunless polar winter, the quantity of ozone—normally created by the action of sunlight on oxygen molecules—could actually increase. What was going on up there?

In his search for explanations, Boville joined a circle of scientists in Montreal, the Arctic Meteorology Research Group, and propelled the group into ozone research using data from instrument-laden balloons capable of rising 35 kilometres and from rockets reaching 70 kilometres....

In 1979 Boville joined the Geneva-based World Meteorological Organization where he set up international programs to monitor both the greenhouse effect and ozone depletion. As one of a handful of ozone experts around the world...Boville was attuned to the theories emerging in the early 1970s that CFCs ascend to the stratosphere where they destroy the ozone that protects life from ultraviolet radiation....Boville's concern prompted him to join the United Nations Environment Programme as ozone specialist, and led him to foster the dialogue between scientists and politicians that culminated in the Montreal Protocol.

Barney Boville has seen a lot of weather and loved it all...he has studied the arctic skies as well as the tropical El Niño ocean warming that upsets atmospheric conditions throughout the world. So where, in his view, is the world's most interesting weather? "I'd have to say the most exciting and changeable weather I ever experienced," he reflects fondly, "was during my 4 1/2 years in Newfoundland."

Analysis:

9. How did Byron Boville's involvement in ozone research expand beyond a purely scientific interest?

10. Do you think that scientists have a social responsibility to report and act upon their field of research? Explain your answer.

Life Under the Ozone Hole

by Brook Larmer. *Newsweek*, Dec. 9, 1991. © 1991 Newsweek Inc.

"Walter Ulloa had no idea he lived in the Twilight Zone. He was too busy herding cattle on the southernmost spit of land in South America to have heard about the blind salmon caught in Tierra del Fuego. Or the pack of rabbits so myopic that hunters plucked them up by their ears. Or the thousands of sheep blinded by temporary cataracts. None of that much mattered—until unusual things started happening to Ulloa himself. After long days on an upper pasture, the 28-year-old ranch hand found that his arms burned "like boiling water," he said. His eyes, swollen and irritated, clouded over; his left one is now completely blind. Another ranch hand was also affected: focusing on objects now makes him weep uncontrollably.

...Every year from late August to early December, the hole over Antarctica expands northward. Its outer reaches now cover the tip of South America, exposing flora, fauna and people like Ulloa to increased doses of ultraviolet radiation. Several Chilean scientists estimate that levels of the carcinogenic ultraviolet-B radiation jumped more than 1,000 percent in Punta Arenas on peak days last year. "This has never happened to a human population at any time in history," says Jaime Abarca, the region's lone dermatologist. "It's as if Martians had landed."

Nothing is more troubling than the possible effects of UV-B radiation on human beings: skin cancer, cataracts, even a weakened immune system. "It's like AIDS from the sky," says Bedrich Magas, an electrical engineer who has waged an aggressive campaign to attract public interest—and research funds. "We get one dose every spring."

Sceptics say Magas is acting less like Galileo than Chicken Little. To date, very little direct evidence exists of a link between the freak occurrences in southern Chile and excess UV radiation.

...Abarca, the dermatologist, has noticed a high incidence of melanomic skin cancer (four times its average rate) and an increase in cases of skin burns and blotches. But he is cautious about blaming ultraviolet radiation. "We need more instruments and information," Abarca says. Neither is forthcoming. A three-year University of Chile project to study the effects of the ozone hole has foundered. The team of scientists has been unable to raise the last $11,000 needed to buy a spectral radiometer, which measures radiation levels. "We are paying the bill for industrialized countries that are depleting the ozone," says Rodolfo Mansilla, who lost 80 sheep to blindness last year. "They have to take an interest."

...Meanwhile, back in Punta Arenas, sunglasses are selling briskly and sunscreen is being stocked in pharmacies for the first time. It is a cold and cloudy afternoon, but Lionel Morales is wearing sunglasses, a wide-brim hat and sun block (factor 15). "I may look silly," he said, "but at least I'm safe." In the Twilight Zone, looking silly may be the only sensible thing to do."

Analysis:

11. *Try to imagine how you might feel if you were born or lived in this part of the world. Describe your feelings in your journal.*

Last Thoughts

Although our understanding of the processes causing ozone depletion are improving, there is little hope of stopping or reversing these processes quickly. Because CFCs remain chemically stable until they reach the stratosphere, and because they take a long time to get there, significant reduction in CFC production today would not show any benefits for many years. The search for alternatives to ozone-destroying chemicals may be extremely costly, yet do we really have any choice but to continue this search?

Imagine!

"CFC production and use in the U.S. alone involves 5,000 companies at 375,000 locations, employs more than 700,000 people, and produces goods and services worth $28 billion."
The State of Canada's Environment, 1991

Students in Action

1. Avoid using aerosol sprays such as hair sprays and air fresheners, especially those containing CFCs.
2. Avoid using air conditioners—use a fan instead.
3. Write to companies that manufacture products using CFCs, e.g. refrigerators, air conditioners, styrofoam; explain your concerns and encourage them to find alternatives that will not damage the environment.
4. Note the daily UV reading for a week or two, and attempt to understand the meaning of these readings.

253

CHAPTER 16
Global Warming

A growing scientific consensus exists that sometime in the next century, the surface of the earth will become warmer than it has been any time in human history.

(John Frior, Deputy Director, National
Center for Atmospheric Research)

Introduction

To those of us living beside the cold Labrador current, the gradual warming of our environment might be thought of as an improvement. However, for the earth as a whole, global warming presents very serious consequences. The melting of many ice-covered environments may cause a rise in the world's oceans, which could make 200 million people homeless as a result of flooding of coastal lowlands. Global food supplies could be at risk as growing conditions change. Even here in Newfoundland and Labrador, there might be some very unpleasant surprises. With changing water temperatures, significant shifts in fish populations could occur. As well, forest ecology, and the industry it supports, might be seriously disrupted.

Dramatic climate change has occurred many times in the history of the earth. For example, the ice ages were related to climate changes, and the demise of the dinosaurs may have been as well. So why are we so concerned about it now? There are several reasons. Widespread human civilization did not exist when major climate alterations occurred in the past. Compared with past changes, the present warming is happening very fast, largely because of this human presence and activity on the planet. In short, this is the first major climate change caused and felt by large human populations.

Although no one can be certain if the climate change we are experiencing is related to natural causes or not, scientists do agree that human activities are contributing to an acceleration in the warming trend. The development of a full understanding of this trend and its implications is, nonetheless, difficult.

Because it is impossible to experiment with climate on a large scale, and equally impossible to reconstruct it in a laboratory, the issue of global warming has stretched environmental science to its limits. It is an issue full of controversy over the rate of climate change and its potential to influence global economics, politics, social disruption, food production, and plant and animal extinction.

What You Will Learn:

- the basic science behind global warming;
- the major greenhouse gases and their sources;
- the changes in global climate over the past century;
- the major consequences of global warming;
- the computer-modelling predictions of global warming;
- the urgent nature of this issue;
- some proposed measures for counteracting global warming.

Imagine!

The term 'climate change' is not synonomous with global warming. Climate change can imply either warming or cooling temperature conditions. Thus, global warming is only one aspect of climate change.

Imagine!

The earth's seven warmest years since 1881 have been: 1981, 1983, 1987, 1988, 1989, 1990 and 1991. The years 1991 and 1988 were the warmest two years ever recorded.

1. *Although there have been distinctive periods of warming and cooling occurring on this planet for millions of years, what distinguishes these past climate changes from the present one?*

2. *What advantages or disadvantages might global warming have for eastern Canada?*

What is Global Warming?

Global warming is linked with the *greenhouse effect*, which describes a natural process involving the interaction of sunlight and carbon dioxide and other gases in the atmosphere. It works something like the process that warms the inside of a greenhouse—hence the name 'greenhouse effect'.

Sunlight passes through the glass of a greenhouse, warming the plants and soil inside. The light energy is converted to **infrared radiation** and the greenhouse heats up, mainly due to the lack of air circulation. Carbon dioxide, water vapour and other gases act something like the glass of the greenhouse by letting solar radiation pass through the atmosphere practically unimpeded. However, after the solar radiation warms the earth and is converted to infrared radiation, much of it is prevented from escaping through the atmosphere back into space. Thus the earth is warmed further. Without 'greenhouse gases', the surface of the earth would be about 30°C colder than it is. This greenhouse effect has existed for eons, ever since carbon dioxide and other greenhouse gases formed in the atmosphere. Global warming refers to an enhanced greenhouse effect, caused by the relatively recent increase of greenhouse gases in the atmosphere (see Figure 16.1). One of the primary causes of the warming trend is the rise in atmospheric carbon dioxide caused by the high level of fossil fuel combustion that energizes modern society (see Figure 16.2).

In 1990, the United Nations Intergovernmental Panel on Climate Change (IPCC) estimated that the average world temperature will rise by 1.3°C by the year 2020, and 3°C more by 2070. Although these temperature changes may not seem extraordinary, you must realize

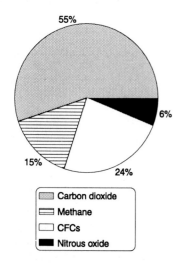

55%

6%

15%

24%

Carbon dioxide
Methane
CFCs
Nitrous oxide

Figure 16.1 - Greenhouse gasses contributing to global warming.

Infrared radiation: heat energy from the sun. The wavelengths have slightly lower energy and are slightly longer than visible light.

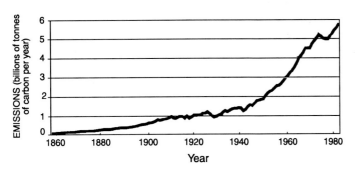

Figure 16.2 - Global carbon dioxide emissions from the burning of fossil fuels.

that they are describing global averages. To put this in perspective, the difference between the global average temperature during the last ice age—when most of this province was buried under a 2 km thick sheet of ice—and today's average is about 4°C.

Another cause for concern is the speed at which today's global climate change seems to be occurring. The present warming trend could be about 100 times faster than the change that followed the last ice age 10,000 years ago. The impact of this rapid change could be disastrous for many plants and animals, which generally need hundreds or thousands of years to evolve and adapt to new conditions.

In order to understand global warming and its influence, it is first necessary to review the carbon cycle (see page 37). Carbon compounds like CO_2 are rapidly moving from one reservoir to another (the major carbon reservoirs are shown in Figure 16.3). Through photosynthesis, plants remove about 120 billion tonnes of CO_2 from the atmosphere each year and put back about the same amount through respiration and decay. This balanced cycle has been disturbed by the excessive burning of fossil fuels, which is adding about 5.6 billion tonnes of extra CO_2 to the atmosphere annually. The oceans, which can normally absorb excessive atmospheric CO_2, are unable to do it fast enough. Hence the build-up of CO_2 in the atmosphere.

On a global scale, Canada is the eighth largest producer of CO_2 emissions, with the United States and the

Canadian Sources of Major Human-Related Greenhouse Gases

Carbon dioxide
- Oil burning - 47%
- Natural gas burning - 30%
- Coal burning - 21%
- Cement/lime production - 2%

Methane
- Landfills - 48%
- Livestock - 34%
- Natural gas leaks - 10%
- Coal mining - 5%
- Other - 3%

Nitrous oxide
- Fuel combustion - 48%
- Chemical production - 27%
- Fertilizer use - 24%
- Other - 1%

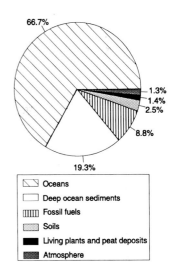

Figure 16.3 - Major carbon reservoirs.

Commonwealth of Independent States ranked first and second. However, on a per capita basis, we are the second largest producer. This is largely due to the high levels of energy needed to cope with our cold climate, and the energy needs of a transportation industry for such a large country.

Analysis:

3. *What is global warming? Use a common everyday example (other than the one in the text) to explain this effect.*

4. *After reading the description of global warming in this text, a friend of yours says, "I am not going to worry about the consequences of global warming because I will not be around when the planet experiences the major effects." How might you respond to this comment?*

5. *Study the carbon cycle on pages 37-39. Redraw the cycle, but this time include diagrams that show human influence in the cycle.*

6. *Canada is the eighth largest producer of carbon dioxide emissions in the world. Suggest three ways that you and your family can help reduce carbon dioxide emissions.*

Predicting the Future Climate with Computers

Feedback Mechanisms

Feedback mechanisms can be either positive or negative. Positive feedback occurs when an event causes a sequence of events that build upon the effect of the first one. For example, if global warming causes large areas of ice and snow to melt, the newly exposed areas will absorb greater solar energy than the reflective snow and ice, so the warming effect would multiply. Negative feedback occurs when one event causes a counteractive effect in some other part of the environment. For example, global warming will likely increase evaporation, which would cause an increase in cloud cover. This would reduce the amount of radiation reaching the earth, thereby causing a cooling effect.

The major difference between weather forecasting and climate forecasting is that the former tries to predict detailed patterns, while the latter monitors global trends and makes general predictions based on them. The main tool that experts use in this process is *computer modelling*.

Computer modelling involves feeding enormous quantities of data into very sophisticated computers that are programmed to evaluate the data and predict the outcomes of changes. This method only began to yield usable results for weather prognosis around 1970. The prediction of future climates using computer models has been occurring for less than 15 years, but it has now become the standard means for trying to understand the earths's complex climatic interactions. However, a number of different models exist, and they do not all agree on the predicted rates of global warming or its precise results.

The difficulty in making accurate predictions about the future of our climate is that it involves so many interconnected factors. These are known as *feedback mechanisms* (see margin). There are thousands of feedback mechanisms affecting climate, many of which are very difficult to quantify. Hence the variation in predictions from the computer models. Other factors affecting predictions are the kinds of assumptions that are made in forecasting and the degree of approximation used.

Computer models are formulated based on the creation of different scenarios. For example, estimates are made regarding expected population numbers, consumption patterns, conservation efforts, etc. Different data input will yield different scenarios, some more optimistic than others. However, it is important to note that even though computer models may differ in their specific forecasts, there is agreement that temperatures will increase, and that the amount and rate of that increase will depend on our concerted ability to reduce carbon dioxide emissions and other greenhouse gases.

Imagine!

Since trees absorb large amounts of CO_2, global patterns of deforestation are a major concern.

The Consequences

The potential consequences of global warming are numerous. For example, sea-level is expected to rise a metre or more over the next century, primarily as a result of the expansion of ocean water as it warms up. Such a rise would flood many deltas and low-lying coastal areas occupied by millions of people. Up to 18% of Bangladesh could be flooded by the year 2050. Also, many of the small oceanic islands might become uninhabitable, and freshwater supplies in many areas could become contaminated by salt. In Canada, it is predicted that we will experience major shifts in wind and rain patterns. The models also suggest a potential temperature increase of up to 10°C in some high latitude areas of this country.

Plants and animals that have taken thousands of years to adapt to the present climate will likely have difficulty adapting to the new conditions. With every 1°C increase in temperature, plant communities would need to move about 90 km towards the poles to survive. Since many plants cannot spread at this rate, they might die out, as would many animal species that rely on them for food. Ocean food chains could be disrupted because rising temperatures may kill large quantities of plankton,

Potential Effects on Agriculture

Although the exact details are difficult to predict, agriculture will probably be disrupted as growing conditions change. The United States will likely be able to continue supplying most of its own food, but exports to other countries are expected to drop by 70%. Growing conditions for agricultural crops may improve in some parts of Canada, Britain, the Netherlands, Scandinavia and the Ukraine; while other areas like Greece, Italy, France, Germany and many developing nations are expected to be hurt. Although these trends might suggest that major agricultural areas could shift northward, this is not likely to happen because northern soils are less productive than those in the south.

Imagine!

All areas will not change with equal speed. Areas between the latitudes of 60° and 90° are expected to change the fastest.

Consequences in the North

Throughout the north, where a permanently frozen layer of ground, called *permafrost*, lies a few metres below the earth's surface, a major disruption is expected, especially along its southern limits. The warming climate is expected to thaw the permafrost, which would disrupt northern ecosystems and create unstable ground for roads, buildings, and pipelines.

Imagine!

People will want to move as conditions worsen in some heavily populated areas, but there will be few places to move to because of already crowded conditions.

the basis for most marine life. Fish populations and distribution could also change in response to altered food and temperature patterns.

Most computer models predict that forests, and as a result the forest industry, will change substantially. For example, they suggest that the climate of most areas of western Canada, which are now occupied by boreal forest, could become more like that of grasslands. Forests are not likely to migrate northward fast enough to match the climate change, so the southern margins of northern forests could die off. A significant expansion of northern forests into the tundra is not expected because of the limitations posed by the poorer soils. The stresses to the forest ecosystem as a result of climate change are in addition to those caused by acid rain and other air pollution.

Supplies of fresh water are expected to shrink as precipitation patterns change and warmer conditions increase the rate of evaporation. Southern Canada, where most people in this country live, may be severely affected by this loss. A reduced water supply would also lead to an increase in the concentration of polluting substances near industrial areas, i.e. in the Great Lakes. It would also influence this country's ability to produce hydroelectric power.

In addition to general warming, experts are predicting that precipitation will vary more throughout the year as well as between years. This may lead to extreme droughts in some areas, even though average precipitation is expected to increase.

Although the consequences of accelerated global warming could be dramatic, it is important to understand that they are predictions—'best guesses' made by experts, based on past trends and sophisticated computer modelling. However, ecosystems, energy flow, nutrient cycles, and climate change are enormously complex, with interactions so numerous and so superficially understood that no one can be sure of the future.

Analysis:

7. *Describe three major potential consequences of global warming.*

8. *How do forest ecosystems normally adapt to climate changes, and why may they not be able to adapt to this current climate change?*

"What we are doing is irreversible. Climate change is on. We don't know how much or when, and we don't know what's tolerable. We do know there's a delay between increasing greenhouse gas concentrations and climate change, and we know climate change will be with us for decades, if not centuries."

Bert Bolin
Intergovernmental Panel on
Climate Change

"If the North is really serious about coming to grips with global warming—whether caused by higher levels of fossil-fuel use or faster rates of deforestation—then debt and unequal trade must be tackled first. Both are reflections of the deep rift between rich and poor which frustrates our search for environmentally sustainable development."

Dr. Vandana Shiva
Research Foundation for Science
and Ecology

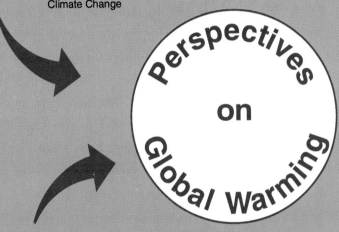

"The developed world might have to invest hundreds of billions of dollars every year for many decades, both at home and in financial and technical assistance to developing nations, to achieve a stabilized and sustainable world. It is easy to be pessimistic about the prospects for an international initiative of this scale, but not long ago a massive disengagement of NATO and Warsaw Pact forces in Europe also seemed inconceivable... Perhaps the resources such an agreement would free and the model of international cooperation it would provide could open the way to a world in which the greenhouse century exists only in the microchips of a supercomputer."

Stephen Schneider
National Centre for Atmospheric Research

"There is no need to panic or over-react to the greenhouse issue. We must keep rebutting the simplistic and often self-interested assertions of those who blame coal as the main cause of the problem."

Peter Cook
Minister of Federal Resource,
Government of Australia

In the News...

The following article extracts represent two somewhat extreme points of view related to global warming. As you read them through, ask yourself the question, "How much of the solution depends on science, and how much on common sense and good will?"

Extract from

The Benefits of Dirty Air

by Sharon Begley. *Newsweek*, February 3, 1992. © 1992 Newsweek Inc.
All rights reserved. Reprinted by permission.

"...Pollutants that cause acid rain got nothing but bad press in the 1980s. Now...atmospheric researchers from seven universities and federal agencies report that these same particles may help ward off a more serious environmental threat: global warming.

The pollutants are sulphate aerosols, tiny particles that help form acid rain and are also unhealthy to breathe. They come from smelting metal and burning fossil fuels such as coal and oil. They reflect sunshine and seed clouds, which in turn bounce sunlight back to space. In both cases the result is a colder planet. That much had been known for years. But according to the latest research, this cooling just about equals the heating effect from "greenhouse gases" such as carbon dioxide (CO_2) and so "has likely offset global greenhouse warming to a substantial degree." In other words, as a result of sulphates there has been less warming than simple greenhouse models predicted.

Last week's report follows another discovery: that the same sulphates may be the planet's best defense against the vanishing ozone layer. In December, Shaw Liu of the National Oceanic and Atmospheric Administration suggested why the thinning ozone layer...hasn't given us all melanoma yet. Writing in Geophysical Research Letters, he noted that the tiny sulphate particles scatter ultraviolet radiation. That's why, despite a 5 percent loss of wintertime ozone in the Northern Hemisphere in the last decade, there is not much more UV reaching the ground, he concludes. Sulphates act like little lead umbrellas.

If all these interactions among pollutants and the atmosphere sound complicated, that's because they are. It was just last autumn that the first complexity got pinned down: a diminishing ozone layer indirectly cools the Earth, and so partially counters greenhouse warming (*Newsweek*, Nov. 4). While we might get cataracts and skin cancer from the extra UV streaming in, we won't have to dike Manhattan quite yet."

Analysis:

9. Explain the irony in this article.

10. What are the possible risks of living under the assumption that air pollutants "offset global greenhouse warming to a substantial degree"?

Extract from

Can Science Save Us?

by Fred Pearce. *New Internationalist*, April 1990. Reprinted with permission.

"...It seems every scientist is chasing a 'technical fix' to shut down global warming. Last summer a Swiss physicist published details of how a giant aluminum mirror put into space could reflect some of the sun's heat away from the earth's surface, so keeping us cool...

Somebody else said it would be easier to reflect the sun's heat back from the surface of the earth. Trillions of white polystyrene balls floating on the world's oceans might do it. Or we could paint the deserts white.

Geologists take a different tack. There is (very expensive) technology available to remove CO_2 from the chimneys of power stations and convert it into liquid carbon dioxide. We could then bury the liquid into old coal mines or oil wells or pour it into sinking ocean currents that would take our pollution into the sea bottom. Scientists have drawn up maps of vast networks of pipelines spreading from the world's industrial centres to coastlines in order to accomplish this task. It might double the price of electricity, they say.

Biologists get in on the act too. One of nature's own methods of recycling CO_2 from the air is to absorb it into the oceans, where it feeds the growth of algae. Up to half the carbon dioxide that we put into the atmosphere today quickly ends up in the oceans. Why not help the process along? In some parts of the ocean, the growth of algae seems to be limited by the amount of iron in the water. So we could sprinkle iron filings on the sea and see what happens....

Amid all the talk of 'technical fixes' we forget that the most obvious methods of responding to the greenhouse effect are probably the best and cheapest....All the technology for saving energy already exists; it just needs applying. It could be brought in on schedule, with no big changes in our lifestyle—and at no real cost. But we could go even further. Why not spend the next 15 years investing in even better energy technologies; there's lots more potential in geothermal, solar and wind power.

And if we were really feeling brave, we could start thinking about bigger changes to the way we live. Like finding ways to allow people to live close to their work (thus cutting down the need for CO_2-producing cars). By comparison with grandiose scientific cures all this seems very dull. Afterwards we would probably say: 'What was all the fuss about? Why didn't we do it before?'

But it always takes a shock to change things. Only the great London smog of 1952 (when more than 4,000 people died within a week from air pollution) jolted the UK to introduce clean-air laws. Now, perhaps a few scary stories about the greenhouse effect may encourage us to do some extremely obvious things about the way we use energy."

Analysis:

11. Review each of the suggested "technical fixes" to counter global warming. What are their strengths and weaknesses?

12. Write a short essay addressing the same question posed in this article: "Can science save us?"

13. What do you feel are realistic options for dealing with global warming?

Last Thoughts

Can We Afford Not to Act?

Climatological experts predict that even if we successfully reduce the emission of greenhouse gases immediately, we will not be able to stop the doubling of CO_2 in the atmosphere. However, if we do not reduce these emissions, the CO_2 level will likely go well beyond that. We have little choice but to prevent temperatures from rising too far and too fast, especially considering that the rate of change is probably more hazardous than the change itself.

In dealing with the problem of global warming, it appears that we have three choices: we can do nothing and hope that living organisms can adapt to climate changes as they occur; we can plan and modify our activities to anticipate the changes; and we can try to reduce the emission of greenhouse gases.

Experts agree that the first of these choices would likely lead to profound ecological, social and economic disruption. Action is needed now to prevent the risks from becoming any greater.

The second choice would include actions such as designing major construction projects with climate change in mind. Major tree planting efforts could also help significantly, providing shade and a cooling effect. Widespread reforestation could help connect separated patches of forest and provide avenues for plants and animals to migrate as conditions change.

Most countries also agree that we must reduce greenhouse gases. Even though the industrialization of developing countries may offset these efforts, most industrialized countries, including Canada, are acting to stabilize or reduce the production of greenhouse gases. Perhaps the most significant way of doing this is through energy conservation.

Students in Action

- Whenever possible, walk or ride a bike instead of driving.
- Help to insulate your home as well as possible.
- Put on a sweater instead of turning up the heat.
- Turn off lights when they are not needed.
- Help your school develop an energy conservation program.

These are very simple actions within everyone's power, but if we all make them a habit, they could make a real difference. We can also ask our political and institutional leaders to promote actions to reduce greenhouse gases (e.g. research and development of non-polluting alternate energy sources—see Chapter 18).

CHAPTER 17
Ocean Pollution

Large surface areas in the mid-ocean as well as near the continental shores on both sides were visibly polluted by human activity. It was unpleasant to dip our tooth brushes into the sea. Once the water was too dirty to wash our dishes in.

(Thor Heyerdal (1950). Heyerdal crossed the Atlantic in a primitive vessel made of reeds.)

Introduction

Scientists estimated that in the first 10 weeks of 1990, 25,000 seabirds were killed by oil in the waters around Newfoundland and Labrador. These deaths were not caused by a major oil spill. Rather, they were caused by passing ships which intentionally flushed out their oily bilge water into the ocean. An oil spot the size of a one dollar coin is enough to kill a seabird.

Endangered sea turtles can swallow as many as 25 plastic bags per day, mistaking them for one of their favourite foods, jelly fish. These plastic bags will eventually block their digestive system and kill them.

The tissues of the white beluga whales of the St. Lawrence River are so full of poisonous chemicals that when they die and wash up on the beach they are considered to be toxic waste. These chemicals have come from the wastes dumped into the St. Lawrence River by the cities and industries upstream.

These are examples of problems occurring throughout the oceans of the world. Our oceans are in trouble!

Understanding Ocean Pollution

Imagine!

A massive algal slick in the waters between Denmark, Sweden and Norway became known as the 'marine Chernobyl' because of its destructive impact on ocean life and coastal areas during the summer of 1988.

Ocean pollutants: substances that enter the sea which lead to a harmful change such as dirtiness, impurity, unhealthiness or hazard.

Because of the earth's water cycle and air currents (see page 35), most of the contaminants that we produce on land eventually reach the sea. Nearly half of all **ocean pollutants** enter as a result of direct dumping or through the rivers that flow through the world's major industrial centres. Another third of the contaminants reach the sea as fallout from air pollution. Lesser amounts arise from activities, such as shipping and offshore oil production.

Once in the ocean, some pollutants accumulate in the tissues of living organisms, build up in the bottom sediments, wash ashore, or remain suspended in the water. The precise effects of the many different pollutants are not well understood, but we do know that, because all the world's oceans are connected by tides and currents, no ocean escapes pollution.

Analysis:

1. *Describe three ways in which pollutants enter the oceans. From Table 17.1, select an example of a pollutant that enters the ocean by each of the ways described.*

2. *"If 75% of the earth's surface is ocean, there is no need to worry about ocean pollution." Based on what you have learned so far in this course, how might you respond to this comment?*

Common Ocean Pollutants

The many types of ocean pollutants, their sources, and effects are summarized in Table 17.1. Oil is one of the most common ocean pollutants. It is estimated that two to five million tonnes of oil contaminate the seas each year. Of this amount, 56% comes from industrial and motor oils, 20% from normal shipping operations, 15% or more from natural seepage and 4% from offshore oil production. Although it is illegal for ships to flush their bilges into the ocean, few ports have facilities to receive and process oily bilge water, so the worst areas tend to be the main shipping lanes. Occasionally a major accident occurs, causing the release of immense quantities of oil. For example, in March of 1989 the *Exxon Valdez* ran aground off the coast of Alaska and released 45,000 tonnes of crude oil into the ocean (see page 274, Alaska's Big Spill).

Grey seal tangled in trawl web on Sable Island

We dump into the oceans many of the offending materials that we need to dispose of on land, such as the material dredged up from the bottom of harbours and rivers to keep shipping lanes open. This material often contains toxic substances and heavy metals that have accumulated from our industrial processes for decades. Five million cubic metres of such materials were dumped in Canadian waters in 1989. In New York City, treated *sewage sludge*, which is the concentrated product of everything that runs through the city's sewers, is placed on barges, then towed out to sea and dumped. Scientific understanding of the impact of this type of dumping is very limited. According to Environment Canada (1991) "...monitoring programs have not been adequate to fully measure short- and long-term effects."

Sea gull trapped by six-pack yoke

Tonnes of waste chemicals also enter the oceans every year. The Mississippi River now carries 10,000 kilograms of pesticides into the Gulf of Mexico annually.

Category	Pollutant	Source	Effect
Biological agents	Sewage	Cities and towns Most Industries	Spreads disease-causing bacteria and viruses; contaminates beaches and swimming areas; causes over-fertilization and algal blooms; depletes oxygen; contaminates food; contains many toxic metals and other substances.
	Animal & Plant Wastes	Household waste Agriculture Industry Abattoirs Forest Industry Food Processing Industry Landfills	Spreads disease-causing bacteria and viruses; contaminates beaches and swimming areas; causes over-fertilization and algal blooms; depletes oxygen; contaminates food.
Dissolved Chemicals and Compounds	Pesticides & Herbicides	Agricultural Industry Forest Industry Household waste	Accumulates in fish and other animals affecting their reproduction; builds up in bottom sediments.
	Metals	Manufacturing and Mining Industries Landfills Natural soil run-off	Poisons animal and plant life; corrodes equipment.
	Fertilizers and Detergents	Agricultural Industry Household waste	Overloads water with nitrogen and phosphorus causing algal blooms; decomposition of excess algae increases bacteria and depletes oxygen.
Undissolved Chemicals, Sediment	Oil	Ships Drill Rigs Service Stations Landfill Sites Household waste Run-off from city streets Oil changes from vehicles	Suffocates aquatic life; destroys thermal protection of fur and feathers; kills fish, birds, mammals and reptiles after ingestion; becomes tarry and destroys bottom habitat of fish, shellfish, worms, etc.; contaminates beaches and swimming areas.
	Metals	Manufacturing and Mining Industries Landfills Natural run-off	Poisons animal and plant life; corrodes equipment.
	PCBs	Transformers and capacitors Some insecticides and paint	Byproducts such as dioxins and furans poison animal life; can cause cancer in people.
	Soil & Silt	Natural Erosion Erosion from Deforestation Agricultural run-off Construction Industry	Fills stream channels; suffocates fish spawning areas; clogs gills of fish and shellfish.
	Plastic	Many Industries Household waste Cargo and other ships	Snags wildlife; some animals mistake it for food; contaminates shoreline.
Temperature	Warmed discharges	Thermoelectric and nuclear power plants; Changing flow of major rivers	Lowers dissolved oxygen in water; changes fish and shellfish habitat; large scale changes in river can influence climate.
	Cooled discharges	Manufacturing Industry Food Processing Industry	Changes fish and shellfish habitat.

Table 17.1 - Sources and effects of major ocean pollutants.

268

Highly toxic chemicals, such as pesticide residues and even warfare agents, have been sealed in drums and dropped into the sea. When these drums eventually deteriorate, the chemicals are released. Among the worst of the metal pollutants is the 8,000 to 10,000 tonnes of mercury entering the ocean each year. In the 1950s, mercury was implicated in causing the deaths of about 400 Japanese and brain damage in about 2,000 more people.

Material causing biochemical oxygen demand (BOD) and total suspended solids (TSS) (described in Chapter 10) are serious marine pollutants. They come from the sewage waste, food processing plants, and pulp and paper mills of coastal towns and cities. Material causing BOD can lead to oxygen starvation at the ocean bottom, while TSS coat the bottom flora and fauna, suffocating life. Many beaches have been closed because of high levels of bacteria in these wastes.

The effects of material with high BOD are magnified by nitrogen and phosphorous compounds contained in sewage and agricultural runoff. These nutrients contribute to major algal blooms which can trigger a number of secondary impacts. Since some of the algae can produce their own toxins, these blooms have caused the deaths of millions of fish and devastated some coastal areas in Scandinavia, the Mediterranean Sea, and Japan. They can also contaminate edible shellfish (see page 214).

Discharges from electrical generating plants, especially thermonuclear plants, have created a **thermal pollution** problem that results in a temperature change of the surrounding water. Since many cold-blooded organisms survive within a narrow temperature range, slight changes in water temperature can result in **die-off** or changes in the timing of events such as reproduction.

Imagine!
About 2 million seabirds and 100,000 marine mammals die each year from ingesting or becoming entangled in plastic.

Nuclear Waste Dumping
Until the early 1970s, Great Britain and the United States sealed low-level radioactive wastes in metal drums and dumped them in the ocean. Since the drums would have deteriorated by now, there is little doubt that animals near the dump sites are accumulating radioactivity. Public protests were influential at stopping the activity in these countries, but Switzerland, France, Japan, and Belgium have continued the practice.

Thermal pollution: a harmful discharge of a warmer or colder substance, usually water, which is significant enough to raise or lower the temperature of a river, lake or ocean in the area of the discharge.

Die-off: the widespread death of many organisms over a relatively short time.

Analysis:

3. *Do you think it is fair to place all of the blame on the ship operators for pumping their bilges at sea? Support your answer.*

4. *What might be the result of releasing waste products causing excessive BOD into a bay where shellfish aquaculture was occurring?*

5. *Thermal pollution is considered fatal to many species of aquatic animals and plants. How might thermal pollutants like warm water be put to better use?*

Different Zones...Different Impacts

Not all ocean zones are affected by pollution in the same way (review the major ocean zones shown in Figure 2.3, page 62). The surface of the oceans is one of the most complex and extensive ecosystems in the world. This thin layer, where water and air meet, is full of plankton and other life. The plankton contains thousands of species of microscopic plants and animals, as well as the eggs of cod, sole, flounder, halibut and other species. Whales, seabirds and fish skim through this zone, feeding on its rich supply of plankton.

The surface layer of the ocean is very sensitive to pollution because many chemicals accumulate in the plankton that live there. This includes atmospheric pollutants that fall on the surface. Floating oil has devastating effects on fish eggs and animals that feed or swim at the surface. A polluted surface layer can poison a large portion of the ocean's complex food web.

Surface pollutants are carried from the open ocean by winds, tides and currents until they are eventually stopped by land and wash up on the shores. Inlets and bays trap and concentrate much of this floating pollution.

The impact of this coastal pollution can be severe, especially at the mouth of a river where fresh and salt water mix (see Figure 17.1). This zone is called an *estuary*. Because it is a place where two ecosystems meet, an estuary can have unusually large populations of plants and animals. Salt marshes, which are shallow areas flooded by the tides, often develop in estuaries. The salt marsh is one of the most productive ecosystems on earth, producing up to three times as much plant material as our best agricultural lands. Shellfish, such as shrimp and clams, are especially numerous in these areas. It is estimated that about 70% of commercial fish species spend part of their life cycle in a salt marsh.

Because of their geographic location, estuaries are subject to serious pollution. The pollutants carried by rivers—such as sewage, industrial waste, silts and oil—pass through the estuaries on their way to the ocean. Many large cities are located at the mouths of rivers, resulting in an additional load of pollutants. When this is combined with the pollutants carried into estuaries by

---

Imagine!

About half of the Baltic Sea's deep waters are dead as a result of oxygen starvation.

...the open ocean is still relatively clean. Low levels of lead, synthetic organic compounds and artificial radionuclides, though widely detectable, are biologically insignificant. Oil slicks and litter are common along sea lanes, but are, at present, of minor consequence to communities of organisms living in open sea waters. In contrast to the open ocean, the margins of the sea are affected by man almost everywhere, and encroachment on coastal areas continues worldwide.

United Nations Environment
Programme Report, 1992

Imagine!

Pollution is not evenly distributed in the sea, but neither are marine animals. Much of the ocean pollution concentrates in areas where currents and upwelling mix and concentrate nutrients. It is these same areas to which foraging animals are most attracted.

---

the ocean tides and currents, the problem becomes even more serious. Many shellfish harvesting operations around the world have been stopped because of the toxins that have built up in the estuarine water and bottom sediments.

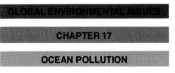
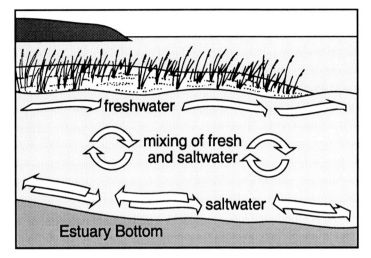

Figure 17.1 - Diagram of estuary showing mixing of water.

___

Imagine!

Over 8 billion kilograms of garbage are dumped onto the ocean every year. A large percentage of this trash is plastic.

___

Analysis:

6. What zone(s) of the ocean are more susceptible to ocean pollution and why?

7. Find out where your community's sewage and other wastes are disposed. Using a topographic map and atlas, determine the route these materials take as they make their way to the nearest ocean.

271

Fixing the Problem

How Bad Is Ocean Pollution?

When a serious event like the break-up of a supertanker pollutes the ocean, we are often horrified by the images we see in the media of its immediate impacts, especially on wildlife. But what does this damage mean?

Some scientists argue that there is little evidence to indicate long-term impacts of ocean pollution on animal populations, except in highly industrialized harbours or near sewage outfalls.

There is little doubt that the oceans have the ability to withstand certain amounts of pollution without showing ill effects. For example, natural oil seepage has occurred for millions of years, and bacteria that can break down this oil exist in the oceans.

The main concern of some is that the quantity and type of pollution entering the oceans has increased substantially through human actions.

Should we be unconcerned about ocean pollution until we have evidence of long-term effects on animal populations, or should we try to prevent potential impacts of our activities? What trade-offs are we prepared to make in order to do this?

Many of the actions needed to deal with other environmental issues, such as ozone depletion, global warming, deforestation, acid rain, and municipal waste, can also help the world's oceans. Yet there are some profound difficulties in correcting certain aspects of ocean pollution. Countries are responsible for, and have the ability to regulate, activities within 200 miles of their shorelines. In Canada's case, this area is so vast that even patrolling it to prevent ships from dumping material is very difficult. Once a substance is dumped and a ship has moved on, it is extremely hard to prove who did it. Meanwhile, shore-based facilities for handling ships' waste are underdeveloped.

Since most of the oceans lie beyond the 200-mile limits controlled by individual nations, marine activity in these international waters is relatively uncontrolled. Also, ocean currents carry pollutants long distances beyond national boundaries. These two factors combined suggest that countries must work together to reduce the problems of ocean pollution.

Many countries still argue that we don't know enough about the effects of ocean contaminants to justify the costs of reducing pollution. Yet some progress is being made. In 1985, the United Nations Environment Programme completed the Montreal Guidelines. These guidelines provide a list of scientific and management principles and strategies to help countries develop action plans to control ocean pollution from land-based sources. In an effort to help reduce ocean dumping of dredged materials and sewage waste, Canada signed an international agreement called the London Dumping Convention in 1990. It requires the phasing out of ocean dumping by 1995. Meanwhile, the impact of pollution from ocean-going vessels is being studied by the international Group of Experts on the Scientific Aspects of Marine Pollution (GESAMP).

"Ironically, deforestation, the principal cause of the loss of biological diversity on land, is also a leading cause of habitat destruction in the sea through runoff and excessive sedimentation."

John Ogden
Florida Institute of
Oceanography

"...disposal of sewage in the sea, done properly, has its supporters. The City of Victoria cites many studies showing that sewage, free of industrial wastes and screened to remove trash, actually benefits the ocean. The seas have been starved of nutrients, the studies indicate, particularly over the last two centuries by the damming of many of the world's major rivers. One study pointed out that the fish catch doubled at a site in the North Sea "due to the stimulating effects of the input of domestic wastes."

Dane Lanken
Canadian Geographic, 1990

Perspectives on Ocean Pollution

"There's an important thing people must realize in planning for a spill this size: No amount of equipment will clean it all up, even if they give you a month's notice to get ready. Look at the expanse of water involved, and figure the time it takes to deploy boats and skimmers and support vessels at 12 knots. Skimmers need barges to collect their oil. Crews need food, ships need fuel, and somebody has to collect the garbage. And nothing works if the weather's bad."

Jim O'Brien
Exxon's water-cleanup
coordinator speaking about the
Exxon Valdez accident, 1990

Ghost Nets
"...many fishing nets are lost or discarded in the oceans—becoming "ghost nets"— and in turn are damaging to wildlife... Ghost nets and net fragments that drift on their own—untended through ocean waters, catching fish and other animals—are extremely hard to find and retrieve. Gill drift nets are designed to be nearly invisible in the water. Sightings of lost nets are rare, but most researchers are convinced that fragments of nets that wash ashore and those found drifting in the sea represent a low percentage of the actual losses."

from *Get the Drift*
Bob Samples, Dr. Cheryl Charles,
and Judy Dawson
Project Wild Supplement

In the News...

The article extracts that follow look not only at some of the effects of pollutants on marine habitats and wildlife populations, but also at some of the real difficulties in reducing and dealing with ocean pollution.

Extract from

Alaska's Big Spill...Can the Wilderness Heal?

by Bryan Hodgson. *National Geographic*, January 1990. Reprinted with permission.

"...the damage had been staggering. Oil had drenched or spattered at least 1,200 miles of shoreline. Experts believed that as many as 100,000 birds had died, including some 150 bald eagles. At least 1,000 sea otters had perished, despite an eight-million-dollar rescue and rehabilitation program. Economic costs had been staggering as well...

Dr. [Jacqueline] Michel has researched the effects of oil spills for the National Science Foundation and NOAA since 1974, participating in studies of the *Amoco Cadiz* disaster, which dumped 68 million gallons of oil on the coast of France in 1978. Within three years scientists found that most of the major impacts had disappeared.

"The story is much the same in all crude-oil spills," she told me. "On exposed rocky beaches with much wave action, little oil is left after a year. On quieter beaches the oil persists from two to three years and is frequently mixed with sand and buried. Salt marshes suffer the most damage, and efforts to clean them are too destructive to do any good. In general, fish and bird populations tend to be replaced. The possible long-term effect on the tidal and intertidal ecosystems will take years to learn."...

"We have tested 5,000 fish sent in by state inspectors and by native subsistence fishers, and so far we have found no crude-oil contamination," I was told by Richard Barrett, head of the state's Division of Environmental Health laboratory at Palmer...

Fish are very efficient at converting hydrocarbons they ingest into metabolites and excreting them from the liver to the gall bladder. Consequently we would expect flesh contamination to be slight...

...Scientists of the Environmental Protection Agency have successfully demonstrated a process called bioremediation, in which oil-eating bacteria native to Prince William Sound, may become the best cleanup crew of all...

Perhaps it's true, I thought, that such catastrophes as the great Alaska oil spill are beyond man's remedy. But are they within man's ability to prevent?"

Analysis:

8. What evidence presented in this article suggests that oil spills may not be as environmentally disastrous as was once thought?

9. Do you think that because of this evidence we should be less concerned about the potential damage of future oil spills than we are now?

Extract from

Montreal starts cleaning up the St. Lawrence

by Dane Lanken. *Canadian Geographic,* June/July 1990. Reprinted with permission.

"The city of Montreal has been doing something over the past two years that it has never done before. It is treating its sewage. At long last, it is taking some of the human wastes, plastic trash and street dirt out of the tremendous quantities of water it borrows from and dumps back daily into the St. Lawrence River.

...As our second largest city, with close to two million people, the quantity of wastes it produces is astronomical. (As much as one million cubic metres of waste water are treated each day, enough to fill about 300 Olympic swimming pools.) Only the enormous volume of the St. Lawrence has allowed the city to flush the problem out of sight for so long and pass the consequences on downstream...

Modern city sewage is almost entirely water. The small fraction that is solid includes human wastes, paper products, personal hygiene items, and in Montreal and other cities where storm water and sanitary sewers are not separate, all manner of street litter: plastic cups and windshield-washer fluid containers, sand, gravel, tennis balls, candy wrappers and cigarette packs.

Sewage also contains a bewildering and horrifying melange of chemicals and poisons. For generations now, discharges from dye-works, print-works, bleach-works, chemical-works, tanneries, breweries, paper-makers, woollen-works, silk-works, iron-works and many others have been poured into municipal sewers. More recently, increasing quantities of acids, oils, solvents, heavy metals and even radioactive materials are being dumped or washed down drains, both legally and illegally, in refineries, factories, foundries and labs.

It is far beyond the capacity of Montreal's new sewage plant, indeed of any sewage plant, to remove chemical poisons from waste water.... As a result, many Canadian cities have been tightening up their sewage discharge ordinances in the past few years. Given some companies' casual attention to such laws, a number of cities have installed sensors in sewers to pinpoint where poisons come from, and most have empowered inspectors to check factory sewer pipes...

But if the province of Quebec is finally getting its sewage act together, Atlantic Canada has not yet reached that stage. Seaside towns and cities there simply pipe their sewage into the ocean, and have only just started to talk about treatment."

Analysis:

10. Does your community dump its untreated sewage into a nearby body of water? If so, find out why this practice continues and if there are any plans in the future to treat your community's sewage.

11. What types of materials were being dumped into the drains in the city of Montreal? What types of materials have you poured down your household drain?

Last Thoughts

Weighing Risk

Although most people are concerned about ocean pollution, there is a range of opinion on what to do about it. Some experts argue that the problems with waste disposal on land are greater than those with disposal in the sea. It's a matter of weighing the risks associated with each choice. Some people suggest that the sea simply represents the best option for certain kinds of waste disposal.

For decades we were unconcerned about what we dumped into the oceans, seeming to rely instead on the dictum: "Out of sight, out of mind." The oceans are so vast and deep that it seemed inconceivable that serious damage could occur. Then, in the 1960s, evidence began to mount that damage was occurring. It became evident that the numbers of people on earth and the products of our activity were sufficient to disturb the interconnected life of the oceans. Slowly it has dawned on us that our perception of the ocean as limitless was a mistake.

The problem of ocean pollution will not be solved in a hurry. The dissolving and diluting power of the immense volume of ocean water means that all parts of the oceans will not be equally polluted; most will actually appear quite clean. However, this should not give us reason to relax our efforts to reduce ocean pollution. Perhaps the biggest change we need is one of attitude. We have to stop thinking of the oceans as limitless pools capable of tolerating infinite abuse. We should also remind ourselves that although the short-term cost of reducing ocean pollution is great, the long-term cost of continuing to pollute is much greater.

Students in Action – making a difference

- Practice the 4 Rs (see Chapter 4, Municipal Waste).
- Snip plastic 6-pack rings and rubber bands used on lobster claws. Sea birds and other marine animals can become entangled in them.
- Clean up a beach, and help keep it clean.
- Take waste oil to a service station.
- If your community has many large ships that come and go, find out if there is a storage system for taking oily bilge from these ships when they dock. If not, write your MHA and MP asking if he/she thinks you have enough vessel traffic to warrant such a system, and if so, why you don't have one.

CHAPTER 18

The Search For and Use of Energy

New generations will need energy, supplied where people live, at a price
they can afford and in forms that make sustainable development possible.

(Advertisement for Statoil, Norway's state oil company)

What You Will Learn:

- the meaning of the terms 'renewable' and 'nonrenewable' as related to energy sources;
- the nature of fossil fuels, including their uses worldwide, and environmental problems associated with their use;
- the connection between energy consumption and the level of a country's development;
- the relationship between energy production and consumption and the environment;
- examples of ways to improve energy efficiency;
- the history and present status of nuclear energy as an alternative source of energy;
- different renewable energy options, their advantages and disadvantages over fossil fuels;
- some of the things you can do to help solve the energy problem.

Imagine!

Only about 10% of the energy used in transportation actually moves vehicles from place to place. The remaining 90% is used up in heat and in overcoming friction and wind resistance.

Imagine!

The change from coal-powered locomotives to diesel locomotives resulted in at least a five-fold increase in fuel efficiency, a major reduction in atmospheric pollution and fewer forest fires.

Introduction

Energy is the most critical of our resources, since it is needed to obtain all other resources and to make them available to society. But modern society's use of energy has increased dramatically since the industrial revolution. In fact, modern economic progress has become intimately tied to increasing energy consumption. For most of the world, the main source of that energy has been and will continue to be for some time fossil fuels—oil, coal and natural gas.

You have already learned about some of the environmental costs of our dependency on fossil fuels—acid rain, global warming, and ocean pollution. In this chapter you will consider global society's use of energy and look at some of the options we have for coping with this high level of consumption.

Energy and Society

Throughout human history, people have derived energy from their own labour, supplemented by the work of domestic animals, and later by water and wind power. Then, in the late 1700s, the Industrial Revolution marked a critical change in our use of energy. First the steam engine, and later the internal combustion engine, created a reliance on high energy consumption. Our dependence on oil came with the internal combustion engine, which could not use coal like the early steam engines. Compared to coal, oil products proved to be cleaner, less bulky, and faster to release their energy, while producing practically no ash. Fuels based on oil became the dominant energy sources by the 1950s and their use has continued to increase since that time. Natural gas, found by itself or in association with oil, has followed a similar but less widespread pattern of use.

The progress of human civilization has been closely linked with the ability to harness energy. In turn, a nation's rate of energy consumption has become an indicator of its standard of living. Generally, energy consumption per person is lowest in the poorest, least developed countries.

As our increasing standard of living demands more and more energy, we find ourselves at a critical point. All the easy oil wells and other sources of energy have been tapped. As global energy consumption continues to grow at its current rate of about 2.2% per year, we have to find and harness more difficult, costly sources.

Analysis:

1. *Make a list of items in your home that require energy to run, and that required energy for their manufacture.*

2. *Choose one item that requires energy to function, and trace that energy back to its original source.*

Types of Energy Sources

All energy sources can be divided into two main groups—renewable and nonrenewable. Renewable energy sources include solar power, biomass, wind and hydroelectric power, which do not rely on limited resources. Nonrenewable sources include fossil fuels, which are not replaced, or are replaced so slowly that they cannot come close to keeping up with consumption. Over time, as fossil fuels become scarcer, their prices will rise and we will be forced to develop new technologies to replace them.

In looking at the main sources of the world's energy (see Figure 18.1), keep in mind that no one source of energy can serve all our needs. For example, coal can not be used directly to run a pickup truck. As long as we have many different ways in which we use energy, we will require a variety of energy sources to suit those needs.

Fossil Fuels

The burning of fossil fuels accounts for 88% of global energy consumption. These fuels are easily transported and can be used to produce other forms of energy such as electricity. Oil provides the main source of energy for the transportation industry, which in turn accounts for the largest drain on the world's oil reserves and contributes to some of our most significant environmental prob-

Imagine!

The cars of the 1990s consume about one-third to one-half as much fuel as the cars of the 1970s. In the U.S., this represents a saving of about five to eight million barrels of crude oil per day.

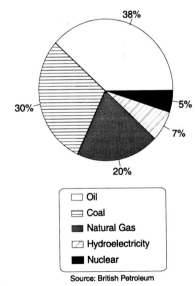

Source: British Petroleum

Figure 18. 1 - Global energy consumption.

The Car Versus Public Transport

There is little doubt that if we could break our infatuation with the automobile, the world would be a better place. In large cities, public transport systems are much more efficient than private cars. A subway can move about 70,000 people past a given point each hour. Compare this with the 8,000 people that can be moved per hour by three lanes of automobiles, each carrying an average of 1.3 people.

Imagine!

There are about 12 million cars on Canada's roads, and we make about one million more each year.

The 1970s Energy Crisis

The energy crisis started when the Organization of Petroleum Exporting Countries (OPEC) temporarily stopped oil exports in order to drive up the price of oil. It went from $2.30 (U.S.) per barrel to $10.50 per barrel. (Later, the Iran/Iraq war of the late 1970s drove the price up even further to about $30 per barrel.) The disruption in supply caused a crisis in the oil-importing countries. In large cities, gasoline was rationed and people had to line up for hours to fill up their cars. It was this event that sparked new interest in energy conservation and the search for alternate energy sources that has continued through the 1980s and 1990s.

lems (see The Car Versus Public Transport in the margin). Since all fossil fuels contribute significantly to global warming, we are forced to try and reduce our reliance on these fuels. Our use of oil, in particular, also has significant economic, political, and environmental consequences.

Since none of the heavy energy consumers—North America, Europe and Japan—has enough oil to meet their needs, they must import it from other countries. It was this reliance on foreign supplies that led to the energy crisis of the 1970s and the turning of a new page in the history of energy consumption (see The 1970s Energy Crisis in the margin).

Whenever political instability occurs in a major oil-producing country, energy supplies are threatened, and the economies of the rest of the world are shaken. This is why some of the military spending in developed nations is directed towards minimizing strife within or between oil-producing nations. They are simply protecting their supplies of oil.

Such military spending is one of the significant hidden costs associated with oil production. Other costs include environmental protection and assistance to companies that find and develop new reserves. Although we use fossil fuels because they are relatively abundant and cheap, they are not as cheap as they seem.

While the oil industry strives to balance oil extraction with the discovery of new reserves, it has not always been successful in this task. The vast majority of the earth's oil sources have already been discovered. New discoveries are lagging behind the rate of consumption, and many are located in environments that make them hard to develop, such as Hibernia off the coast of Newfoundland.

Unlike oil, coal is relatively abundant in North America, but it can substitute for oil in only a few situations, most notably in producing electricity. However, coal burning can be a major contributor to acid precipitation since it can produce significant sulphur and nitrogen oxide emissions. The severity of these emissions can vary greatly depending on the amount of sulphur in the coal and the combustion technology used. As well, strip mining and open pit mining, the main methods of coal extraction, can significantly alter the ecosystem overlying a coal deposit.

Natural gas, which is usually found along with most oil deposits, is a clean-burning fuel whose use is increasing throughout the world. It is usually transported through pipelines, although this is changing, as technological improvements have allowed the liquifying of natural gas and its shipment in specialized tankers.

Imagine!

The U.S. has enough coal reserves to supply its energy needs until well into the 22nd century.

Analysis:

3. *What are the three major types of fossil fuels? Of the three, which has the least impact on the environment?*

4. *How might the conservation of fossil fuels be a better alternative to seeking new sources of these fuels?*

Nuclear Power

Electrical power can be produced from the heat generated by nuclear reactions within radioactive minerals, such as uranium and plutonium. The use of nuclear power experienced rapid worldwide growth beginning in the 1950s, followed by what may have been a slow death in the 1980s and 1990s. Heavy investment began with the belief that nuclear power would be the energy supply of choice by many developed nations. It is favoured by some because it normally produces no air pollution or ash, it is considered safe, and it is affordable.

By the end of 1989, there were 426 operating nuclear reactors supplying about 17% of the world's electricity, and 96 more were being built. The biggest users of this type of energy have been the U.S., France, Germany, Spain, Russia and other Baltic countries, the United Kingdom and Japan. In 1992, Canada had 18 nuclear reactors.

Despite its rapid growth, public interest in nuclear power began to fade in the mid 1970s. No new orders for nuclear power plants have been placed in the U.S. since 1978, and work on many developing plants has stopped. Many people began to believe that the problems associated with nuclear power were not worth the benefits.

The greatest public concern is with the risk of an accident. Although a nuclear power plant will not ex-

Three Mile Island

At Three Mile Island, Pennsylvania, human error caused a potentially disastrous accident in 1979 that nearly resulted in a meltdown. Fortunately, backup safety systems brought the problem under control, preventing the evacuation of 300,000 people from Harrisburg. By 1985, cleanup was still underway and had cost as much as the construction of a new power plant. Safety measures have been significantly upgraded in most nuclear power stations since this event. No one was killed or injured in the accident.

Nuclear Waste Disposal

A more persistent, but less urgent problem with nuclear power is the issue of waste material. After the fuel bundles outlive their usefulness producing electricity, they must be replaced. The old ones remain highly radioactive for hundreds of years, so they are usually stored in highly secure, deep-water pools at the nuclear plant. Most plants continue to accumulate these radioactive fuel bundles, since few countries have established acceptable long-term disposal methods. After about 500 years, the radioactivity is much lower and less hazardous. However, safe storage is still needed for up to 240,000 years.

plode like a nuclear bomb, the reactor core can overheat, which could result in a *meltdown*. The intense heat could cause a steam explosion, which could release radioactive materials into the atmosphere. Although reactors in the western world are built to prevent this type of occurrence, it did happen at Chernobyl in the Ukraine in 1986. Hundreds of thousands of people had to be evacuated from the area, 31 people died from the direct effects of radiation, and many hundred were hospitalized with radiation sickness. Increased levels of radiation were also detected around the globe, the long-term effects of which are still not well understood.

The nuclear reactors in use in Canada, the U.S. and Europe are of a fundamentally different design from the one at Chernobyl, making the possibility of a similar accident quite remote. Nonetheless, there have been accidents at North American plants that were potentially catastrophic (see Three Mile Island on preceding page).

Another problem with nuclear power is that the reactors are wearing out faster than was first anticipated, and they are not producing power as efficiently as they should. As well, once a plant has outlived its usefulness, it is *decommissioned* or closed down—a very expensive process that increases the cost of the power produced during the useful life of the reactor.

Analysis:

5. *What are two factors that may be causing some countries to change their view of nuclear power as the ultimate energy source?*

6. *Through research, find out how the CANDU nuclear reactor works and why it is considered safer than many other models.*

7. *Research the methods of nuclear waste disposal that are currently being investigated.*

Hydroelectric Power

Imagine!

If all existing hydroelectric power were replaced with power from fossil fuels, an additional two billion tonnes of carbon dioxide would be produced each year.

Given that sunlight drives the water cycle, hydroelectric power—which is obtained from the energy of flowing water—is a converted form of solar power. It is a mature technology compared to other forms of renewable energy, that generally requires the building of huge dams to create a large storage area and a redirection of water flow. The redirected water turns generators which produce electrical energy.

Hydroelectricity now provides about 20% of the world's electrical energy, although total generation could increase about six times by the year 2020. It is used most heavily in the developed countries, but its greatest potential lies in the developing countries. For example, North America produces 33% of the world's hydroelectric power using about 59% of its potential sources. In contrast, the Third World has used only about 7% of its potential sources thus far in its production of hydroelectricity.

Although hydroelectricity causes no air pollution, it does cause some significant environmental problems. Large projects can displace communities and seriously alter lifestyles and ecosystems. River systems are also changed radically. Immense areas are flooded upstream, while the water flow downstream can vary considerably as power requirements change. Dams can also trap silt that would otherwise fertilize land downstream, and they can even cause the spread of disease. This happened downstream from Egypt's Aswan High Dam, with the spread of shistosoma, caused by a parasitic worm.

Thousands of caribou were drowned in northern Labrador in 1984. Although unproven, it was strongly suspected that the accident was linked with a surge in water levels caused by a hydroelectric project in Quebec.

Analysis:

8. *Find out more about the James Bay hydroelectric power project. What are some of the environmental and social problems associated with this particular project?*

Solar Power

A bare foot on hot pavement and the interior of a car on a hot day are proof of the everyday heating power of the sun. In fact, the total solar energy reaching the earth each year is equivalent to that produced by 60,000 billion tonnes of oil—about 20,000 times the energy that we now use. The sun's energy heats all regions of the earth, but not necessarily with equal intensity or duration.

Solar energy must be concentrated, converted, stored, and used in appropriate applications. *Passive solar energy* occurs when the flow of energy is by natural means, such as conduction, convection and radiation. It involves no moving parts such as pumps or mechanical equipment. *Active solar energy* uses mechanical equipment to move energy around. A greenhouse might com-

The chief barrier to more widespread use of passable solar design is ignorance.

(Center for Renewable Resources)

Imagine!
Nine out of ten houses in Cyprus have solar panels.

Photovoltaics

Electricity can also be produced directly from sunlight using *photovoltaic cells*. A five-square centimetre cell can produce about the same amount of electrical energy as a standard D-sized flashlight battery. When many cells are linked together, they can produce any amount of electricity. The cost of solar cells has been reduced substantially since they were first developed, allowing banks of cells to be used in power plants. They require no fuel and little maintenance, and they are very reliable, noiseless and non-polluting. Considerable research continues on using photovoltaics to generate hydrogen, which in turn can be used in automobiles.

Imagine!

Between one third and one half of household energy consumption goes towards the heating of water.

Imagine!

Research by building experts in Berkeley, California suggests that $8 million dollars spent on reducing heat loss through windows could save $300 million worth of oil.

---

bine both—with the sun warming the inside passively, while fans move the heated air about.

A relatively new technology involves using lenses or mirrors to focus sunlight on a container of liquid. The heat produced in the container can be used to drive an electric generator. This is known as a *solar-thermal plant*.

Since Israel completed the first solar power station in 1979, many other countries have also begun to produce significant amounts of solar energy. By 1990, about 20 other solar power stations had been built, with about half of them located in California.

The cost of capturing solar energy for everyday use continues to fall to affordable levels. Solar technology is not yet advanced enough to supply major energy needs for large populations, but it can supplement other forms of energy.

Energy Conservation

Energy conservation holds the greatest promise for lowering energy consumption without disrupting economies and living standards. Improved energy efficiency in the U.S. and Canada, for example, was so effective during the late 1970s and early 1980s that their economies continued to grow without a parallel increase in energy consumption.

Standard conservation practices, such as turning off lights and turning down the heat, are important, but in total they are likely to reduce consumption by only a small percentage. Major conservation is achieved through the redesign of our equipment, transportation systems and buildings so they use less energy in the first place and so they capture waste energy wherever possible.

For example, with *superinsulated buildings*, standard insulation is doubled, and walls and roofs are made as airtight as possible. When this is combined with passive solar radiation and perhaps a small amount of energy from an alternate source, homes can be warmed by minor sources of heat, such as light bulbs and refrigerators, thus reducing energy costs by 50% to 80%.

Successful energy conservation has also occurred in the automobile industry. Fuel efficiency has improved through the use of lightweight materials, radial tires, smaller vehicle size, more efficient engines, and improved aerodynamics. However, North Americans are less inclined to purchase fuel-efficient vehicles than Europeans or the Japanese, mainly because our fuel cost is still relatively low.

Energy Around the World

Seventy percent of the world's commercial energy resources are consumed by one fifth of the world's population. Canada uses more energy per person than any other country in the world, mainly because we are a large, cold country that is also relatively wealthy. We also produce and export high quantities of materials such as paper and aluminum, which require large amounts of energy in production. But energy consumption in developing countries is growing, and there is little doubt that the cost to the global environment will be severe.

Take China as an example. It relies heavily on coal, has a huge population, a rapidly growing economy, and low energy efficiency. Globally, it is a growing source of CO_2 and acid precipitation emissions. Clearly, we need to help them develop the most efficient technologies available to help reduce further environmental threat.

Some Comparisons in Energy Consumption

Units represent millions of tonnes of oil equivalents (MTOEs) per year. The figure for Canada is based on 1990 data; all others on 1987 data.

Africa 184
Canada 280
Japan 343
China 629
W. Europe 1182
Former USSR 1240
U.S. 1691

Sources: *State of the Environment Report*, 1991
Atlas of the Environment, 1990

Imagine!

Biomass in the form of wood, crop wastes, dung and sewage is the dominant energy resource in the developing world. In 1992, three billion people in developing countries depended on wood for energy. However, overharvesting of this resource leads to the creation of deserts and topsoil erosion.

Analysis:

9. *Investigate one of the following sources of energy: wind power, geothermal energy, tidal power, wave power or biomass energy. Describe the technology, the present trends in its use, and its advantages and disadvantages.*

10. *If you were choosing between two cars, a slower one with excellent fuel efficiency, and a fast one with poor fuel efficiency, which would you choose? Justify your answer.*

11. *Considering what you now know about global warming and acid precipitation, briefly describe the effects of energy consumption in the Third World reaching levels similar to that of developed nations.*

12. *Find out the prices of gasoline in several different countries and compare them to current prices in Canada. If you lived in a country where the price of gas was much higher than it is here, how would it affect your behaviour? What are the implications of this for the environment?*

"A nuclear power plant was recently constructed at Darlington, Ontario. ...when all
the reactors are finally on stream sometime this year the cost will have ballooned
by more than 500% to $13.4 billion. That's
more than the entire gross domestic product of Nova Scotia...Instead of building the
Darlington plant, Ontario Hydro could
have: given, free of charge, every family in
the province a new, energy-efficient refrig-
erator; insulated and weatherproofed
every house; and installed a high efficiency furnace in every house. This would
have saved more energy than Darlington
has been built to produce..."

Jane Sherwood
Canada and the World, 1991

"Just getting the commercial sector
alone to install the most efficient fluorescent lightbulb can eliminate the
need for the amount of energy provided
by two nuclear reactors or four coal
plants at a price far lower than building
two nuclear plants."

David Suzuki,
1991

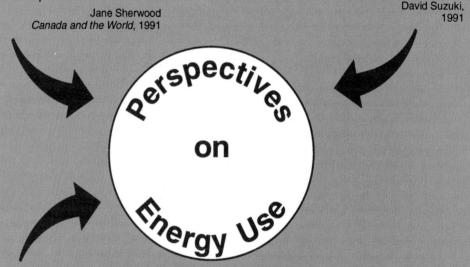

"We could not have flown to the moon on oil or gasoline. We did it on hydrogen. We did it
because hydrogen is a better fuel than oil, just as oil is a better fuel than coal. We always
do the toughest jobs with the newest fuel...The technologies we are moving towards will
mimic nature in an almost eerie way. But we do not have the billion years it took nature to
evolve to where we are now. We have to do it using the one great thing that evolution gave
us—intelligence...and we probably have to do it in less than a century. If we do not, we will
destroy the planet."

David Scott,
1990

In the News...

Our dependence on fossil fuels as a primary energy source cannot continue forever. The following two articles look at some of the alternatives, including some of the drawbacks associated with their adoption.

Extract from

Alternative Energy

By R.C. Mackenzie. *Canada and the World*, Dec. 1991. Reprinted with the permission of Canada and the World magazine, Oakville, Ontario.

Wind - The Unpredictable Source

"...In recent years, the emphasis has been on converting wind energy to electrical energy, but the costs of doing this still cannot compete with hydro power or coal-fired generation of electricity. The cost of wind-generated electricity, however, has declined sufficiently to be competitive with that produced by nuclear power stations.

The problem with using wind power more extensively in Canada is that the windiest parts of the country are along coasts or in the north; areas that are relatively unpopulated. While this does not encourage much interconnection of wind turbines to existing power grids, it does reinforce the notion that wind energy is practical in remote areas.

Hydrogen - Fantasy Fuel Or The Planet's Best Bet?

...When an electric current is passed through water, hydrogen and oxygen are produced. Reverse the procedure by combining hydrogen and oxygen and the result is water and an electric current. Thus, hydrogen is completely renewable and virtually inexhaustible. Burn it in an engine and the principal byproduct is water; no carbon dioxide, no sulphur dioxide, none of the nasties that come from gasoline combustion. And, the energy content of hydrogen per unit of weight is three times that of gasoline, and twice that of natural gas.

There are some problems with hydrogen, though. It usually exists as a gas, and if compressed to a liquid state, any storage tank needed to contain it would be extremely large and heavy. This makes it impractical as a liquid vehicle fuel...Also, the costs of producing hydrogen cannot compete with the low price of fossil fuels in Canada.

It's expected that most of the current obstacles will be overcome, and perhaps, by the turn of the century, hydrogen will take an important place in the world's energy mix."

Analysis:

13. *Which of these two alternatives do you think holds the most potential for our province?*

14. *Find out approximately how much money is being spent by government on the Hibernia offshore oil development. Then investigate some of the other energy sources listed in this chapter. If you had to decide where some of the Hibernia energy dollars should be redirected, what source of energy would you pursue? Explain your answer.*

Extract from

It's a Question of Cost

by R.C. Mackenzie. *Canada and the World*, Dec. 1991. Reprinted
with the permission of Canada and the World magazine, Oakville, Ontario.

"To modern-day Don Quixotes, tilting in favour of windmills rather than against them, it's better to use any energy source rather than fossil fuels or nuclear energy. This is particularly true if the energy source–such as solar and wind–is renewable.

It is easy to agree with this notion when faced with the environmental damage caused by the use of coal and oil, and the fears that surround any nuclear energy project. But replacing these energy sources with alternative energy systems will be difficult. Like it or not, economics plays a crucial role in any changeover, and alternative energy systems have been significantly more expensive than traditional energy sources.

Thanks to improvements in various technologies, the cost of producing energy from alternative sources has come down during the past decade. The cost of oil and gas, however, has gone down even more dramatically, and electricity produced by hydro power or coal burning continues to be a bargain. This makes it difficult for power utilities and individuals to justify a headlong plunge into renewable energy technologies.

Furthermore, reports of the impending demise of fossil fuels continue to be inaccurate, mostly because geologists keep finding new reserves as quickly as the old ones are exhausted. The best guess is that oil and gas will be around for a few more decades and there is enough coal in the ground to last hundreds of years. The abundant supply of these fossil fuels also makes it hard to decide in favour of switching over to renewable energy technologies.

In their favour, however, is the environment-driven political pressure now being brought to bear on fossil fuels. It is easy, for example, to be outraged by the magnitude of the damage done to buildings, lakes and forests by acid rain. And, the burning of fossil fuels is one of the main causes of acid rain.

Energy experts now realize that fossil fuels aren't as cheap as they seem when you add in the cost of the environmental damage they cause. Accounting for fossil fuel damage makes comparisons between them and alternatives fairer and the difference in cost smaller. Even so, Canada is not about to make a major switch to alternative energy sources in the near future. Nonetheless, much progress has been made, particularly in small-scale uses, to which renewable technologies are well-suited."

Analysis:

15. *Why are governments and companies not moving more quickly towards alternative energy sources? What steps could be taken to help the process along?*

16. *What are some of the environmental costs associated with fossil fuel extraction and use?*

Last Thoughts

Generally, it costs more to solve an environmental problem than it does to prevent it in the first place. In spite of this, most governments, including Canada's, are not likely to switch to alternative energy development on a big scale. However, both industry and governments continue to seek solutions. In this search, they ask "Where is our money most wisely spent? What technologies should we pursue?" The solutions seem to rely, not on any one technology, but on a combination of various energy sources and aggressive conservation policies.

Many experts still believe that we must either reduce our dependency on fossil fuels as quickly as possible or face the consequences of accelerated global warming and acid rain. However, one country alone cannot achieve this goal. It will take many countries acting together and strong political will to reduce the world's energy consumption. More importantly, it will require each of us, especially in the western world, to change our patterns of high energy consumption. With lowered expectations in our part of the world, we can set new standards for developing countries as they develop their energy resources.

Imagine!

In Newfoundland and Labrador, many small-scale hydroelectric projects were developed in the 1960s. There has been a resurgence in interest in such projects, as well as in alternative energy, with provincial government's Strategic Plan For Energy Efficiency and Alternative Energy, adopted in 1990. Even these small projects are controversial, since they can alter people's traditional use of rivers and can interfere with wildlife such as salmon.

Students in Action

1. As you reach the age when you consider car ownership, stop and seriously question your need for a car. If you decide that it is a necessity, consider buying the most fuel-efficient one that you can.
2. Reduce, recycle and reuse. All products require energy in their production. The longer you postpone discarding a product, the more energy you save.
3. Let your MHA or MP know that you are concerned about energy consumption.
4. Use fluorescent light wherever possible. Direct the light to the areas you are using rather than the whole room.
5. Help your family improve the energy efficiency of your home by installing weather stripping and water-saving shower heads. To save even more hot water, make your showers as short as possible.

CHAPTER 19
Hazardous Waste

The chemicals to which life is asked to make its adjustment are no longer merely the calcium and silica and copper and all the rest of the minerals washed out of the rocks and carried in rivers to the sea; they are the synthetic creations of man's inventive mind, brewed in his laboratories, and having no counterparts in nature.

(Rachel Carson, author and environmentalist)

Introduction

Based upon a foundation of ecological principles, you have now examined a number of important environmental issues, both in this province and throughout the world. Next you will consider one more global environmental issue—hazardous wastes. But this time you will be in charge of the investigation and analysis yourself.

It is unlikely that all of your future careers will be directly linked to environmental science, but as an adult you will continue to be challenged by issues like those you have encountered in this text. The issues will change and new ones will arise, but if you wish to form valid opinions and act responsibly on these issues, you must learn to consider all perspectives and examine them critically.

In looking at hazardous wastes you will get some guidance, here and in the activities provided by your teacher, on what to investigate and how to approach your research. To get you started, the following provides an outline of the main research areas for this topic.

Hazardous Waste
A Blueprint for Learning

What are Hazardous Wastes?
- The definitions of *hazardous* and *toxic* substances.
- The difficulty in categorizing substances according to these definitions.
- The main types of hazardous wastes.
- Where hazardous substances come from.

What are Some of the Effects of Hazardous Wastes?
- Environmental effects.
- Human health risks.

How do we Manage Hazardous Wastes?
- Management before disposal.
- Disposal management.
- Exporting hazardous waste.
- 'From the cradle to the grave' management.

Researching
Getting at the Issue

There are countless sources to which you can turn for information, not only for environmental issues like hazardous waste, but for any research topic. Listed below are a few ideas to get you started.

- School resource centre or community library
- Newspapers (local, national, and international)
- Government agencies
- Magazines and journals
- People (family, friends, community members)
- Computer networks
- Film, video and audio resources
- Non-government organizations
- Interpretation centres, parks, museums and galleries
- Television and radio
- Public events, seminars and workshops
- Corporations and businesses
- Music, art and literature

See the "Hazardous Waste Learning Unit" provided by your teacher for many more ideas, resources and activities that can help in your study of this topic.

Last Thoughts

An entire universe and just one Earth
Really small compared to the rest
But here we are and here we'll stay

Mandy Pippin

Let's treat Canada as if we plan to stay.

Susan Holtz

You have reached the conclusion not only of this chapter, but also of this course in environmental science. Yet in many ways this represents a beginning rather than an end. You have learned something here about basic ecology and its application to a number of local and global issues. But these issues will change, and new ones will arise. Even our understanding of ecology is incomplete, with much yet to be revealed. As a global citizen, you are embarking on a lifelong journey of learning.

The building of a sustainable society will continue to be an enormous challenge—for individuals, governments, business and industries. But there is cause for real optimism. We are better equipped now with information about our environment than at any other time in history. There is a growing social consciousness that is recognizing environmental problems and is working towards solutions. But it will still not be an easy journey. Along the way, we must constantly look with honesty at ourselves and ask the question, "Am I part of the problem or part of the solution?"

Glossary

Abiotic factors: the nonliving parts of the biosphere such as sunlight, minerals, temperature and water.

Absolute abundance: the actual number of animals or plants in an area in contrast with relative abundance.

Acid rain: rain that is made acidic by air pollution.

Acid precipitation: rain, snow or hail that is made acidic by air pollution.

Acid deposition: fallout of acidic particles from the air; includes both wet and dry fallout and fog.

Acid shock: a relatively sudden increase in acidity levels.

Active solar heating: the use of mechanical means to store and transfer solar energy to areas where it is needed.

Aerial survey: a survey obtained by flying over an area and counting the animals in it.

Air quality: an indication of the amount of pollution in the air.

Air pollution: the presence of impurities in the air.

Allowance: amount of catch allowed the inshore fisheries sector; a portion of the overall quota.

Anadromous fish: fish which spend most of their lives in salt water but move to fresh water to spawn.

Aquaculture: the cultivation of aquatic plants or animals.

Aquatic: associated with water, either fresh or salt.

Aquatic realms: the two water biomes: oceans and freshwater.

Area/quota system: a system of dividing a region into sections or management areas, each of which has certain characteristics slightly different from neighbouring areas. Each area is assigned a harvest quota.

Artificial environment: an environment where most factors influencing the production of plants or animals are controlled.

Atmosphere: the portion of the biosphere containing air.

Bacillus thuringiensis or **B.t.:** one of the most popular biological agents for controlling insects that experience a caterpillar stage.

Biochemical oxygen demanding (BOD): a measure of the level of oxygen consumed in the decomposition of a substance in the environment.

Biodegradable: capable of being decomposed by living matter, especially by bacteria.

Biodiversity: the variety of plants and animals in a given area.

Biomass: mass of one group of living organisms.

Biomes: large portions of the earth with similar climate, soil, plant and animal communities.

Biosphere: the thin skin of the earth's surface and the air above it containing all life. It consists of three layers: the *atmosphere* (air), the *hydrosphere* (water), and the *lithosphere* (rocks and soil).

Biotic factors: the living parts of the biosphere– i.e. plants, animals, bacteria and viruses.

Bloom: the rapid multiplication of organisms caused by a combination of temperature and water.

BOD: biochemical oxygen demand.

Bog: a nutrient-poor peatland.

Boreal forest: northern forest such as those of Newfoundland and Labrador comprised mainly of coniferous trees such as fir and spruce.

Brackish: mixed fresh and salt water.

Buffer: a substance that lessens the impact of another potentially damaging substance.

Bycatch: the accidental catching of different fish or other organisms from those you set out to catch.

Canopy: the uppermost layer of a forest, consisting of interwoven tree tops.

Carbohydrates: an energy-rich substance made from carbon, oxygen and hydrogen — usually a starch or a sugar.

Carnivore: animals that eat herbivores.

Carrying capacity: the maximum number of a species that an area will support for a sustained period.

Catadromous fish: fish, such as eels, which spawn in salt water but spend most of their lives in fresh water.

Cellulose: a sugar compound that forms the walls of plant cells.

CFCs: Chlorofluorocarbons.

Chlorine: a naturally occurring element in the earth and seawater; in gaseous form it is highly toxic.

Chlorofluorocarbons: chemicals used in refrigeration and other industries that contribute to a gradual thinning of the ozone layer.

Chromosome: that part of a cell that contains the genetic code.

Classification survey: a survey to determine the age and sex classification of animals.

Clearcutting: harvesting of all the trees in a large area, leaving extensive open cutovers.

Climax community: the final stage of successional change–a stable community.

Closed system: a system in which there is no, or minimal, input of new material from the outside.

Community: an association of organisms living in a common environment.

Compost: organic wastes such as kitchen scraps mixed with soil in the presence of oxygen.

Computer modelling: a tool used to monitor global trends and make general predictions based on them.

Crustal plates: the basis of the earth's crust.

Cyanidation: a process used to dissolve exposed metal particles.

Daily bag limit: the number of animals a hunter is allowed to harvest or "bag" daily.

Decibel: a unit of measure for sound levels.

Decomposer: bacteria, plants and other animals that feed on dead plant or animal tissue.

Decomposition: the separation of a substance into simpler constituents.

Defoliation: any action that causes the loss of leaves or needles; a pest that causes this type of damage is called a *defoliator*.

Degradable: capable of being reduced or broken down to simpler molecular structure.

Detritus: dead and decaying organic matter.

DFO: the Department of Fisheries and Oceans.

Diarrhetic shellfish poisoning: a type of poisoning in humans that causes diarrhea.

Die-off: the widespread death of many organisms over a relatively short period of time.

Dioxin: a complex organic compound, usually a byproduct of chemical reactions involving high temperatures and chlorine.

Diploid: having two chromosomes.

Dobson unit: a unit for measuring atmospheric ozone.

Drive count: a method of measuring the number of animals in a given area.

Dry deposition: acidic particles in the atmosphere that collect in clouds and eventually fall to earth, not as precipitation, but as dust.

Earth stewardship: the act of looking after the earth as you would your home.

Ecological processes: the relationships among living organisms and with their nonliving environment, including energy flow and water, gas, and mineral cycles.

Ecology: the study of the interrelationships among living things and their nonliving environment.

Ecoregion: a defined area where living organisms share the same basic living conditions.

Ecosystem: a self-supporting community — plants and animals interacting with each other and the non-living environment to produce a balanced system. Since it is an idea rather than a place, it may be as small as a puddle or as large as the whole earth.

Ecotone: areas between neighbouring ecosystems sharing characteristics of each.

Effluent: liquid waste.

Emergents: tall trees that rise above the canopy of a tropical rain forest.

Energy chain: the transfer of energy from one level of an ecosystem on to the next.

Environmental science: the process of applying ecological principles — the natural workings of the planet — to human use of the environment.

Environmental impact statement: a report describing the assessment of the impact of human activities or developments on the environment.

Environmental impact assessment: a study of the impact of human activities or developments on the environment.

Epiphytes: plants that grow upon other plants, but are not generally parasitic.

Estuary: the mouth of a river where fresh and salt water mix.

Extrapolate: to predict an unknown figure based on a sample of known data.

Fen: a peatland influenced by rich ground water.

Fenitrothion: a chemical insecticide used to control many insect pests.

Finfish: salmon, trout, cod, and others.

Fish ranching: a type of fish production where fish are released to the wild then recaptured for harvesting.

Fish habitat: a place that provides all the food, water and space necessary for fish to thrive.

Fish stock: a group of fish that have a common genetic make-up, inhabit a particular region, and maintain a similar migration pattern.

Fisheries moratorium: a cessation of fishing of a particular species or population of a species.

Food web: an interlocking pattern showing the eaters and the eaten.

Fossil fuels: the term used to describe coal, gas and oil.

Fry: young fish fresh from the spawn; young of salmon in second year.

Gaia: the Greek goddess of mother earth. Also the name of a theory to explain the interrelationship of living things with the earth.

Gaseous: having the form of or being gas; related to gas.

Genetic diversity: the differences in genetic makeup of individuals within the same species.

Geochemical: the chemical composition of rock and sediment samples.

Geophysical: the physical properties of rock.

Geophysical anomoly: an unusual physical property in rock.

Geothermal plant: a plant that produces energy by utilizing the heat of the earth's interior.

Grab sample: a chip off outcrops or from surface deposits, analyzed in a laboratory to determine what metals the rock contains.

Green Plan: Canada's long-term plan for making this country one of the most environmentally-friendly nations of the world.

Greenhouse effect: a natural process that warms the earth due to the interaction of sunlight and carbon dioxide and other gases in the atmosphere.

Greenhouse gases: gases with a molecular structure that give them their energy-absorbing char-

acteristics.

Habitat: a place that provides all the food, water, shelter and space necessary for a particular organism to thrive.

Hazardous substance: a substance that is either toxic, will ignite at a relatively low temperature, will explode or is highly reactive.

Heaths: common shrubs of the barrens and tundra that are members of the blueberry family; most have evergreen leaves.

Heathland: alternate name for barrens used in other parts of the world.

Herbivore: animals that live mostly on green plants.

High-grade ore: mineral ore with a high concentration of a specific metal.

Home range: the extent of area an animal needs to survive.

Homeostasis: the maintenance of a stable condition that depends on many interactions in order to remain stable.

Hormones: naturally produced chemicals that control the function of many organs.

Host rock: the rock within which a mineral deposit is found.

Hydroponic: the growing of plants in water and nutrient solution, instead of in soil.

Hydrosphere: the portion of the biosphere containing water.

Incineration: the burning of waste or other material to reduce it to ash.

Infilling: the gradual change from open water to a bog caused by plant life growing in from the edges.

Infrared radiation: heat energy from the sun. The wavelengths have slightly lower energy and are slightly longer than visible light.

Inorganic: material that does not contain carbon and was never living, such as minerals.

Inorganic fertilizer: fertilizer made from non-living or human-made substances.

Inshore fishery: a shore-based fishery conducted mainly during the summer months using hand lines, gill nets and/or cod traps.

Integrated forest management: the management of forest ecosystems considering the variety of roles and uses of a forest.

Integrated pest management: the combination

of practices used to manage forest pests, involving monitoring systems, rules for decision-making, and pest control methods.

Integrated waste management: the use of a variety of strategies and methods for the reduction, safe disposal and general management of waste in our society.

Interdependence: a mutual reliance between species that contributes to the well-being and survival of each.

Joule: the amount of energy expended when a force of one newton is exerted through a distance of one metre.

Leaching: the passing of liquid through material (e.g. water through a dump site) to remove or dissolve soluble matter.

Licence return: the portion of a hunting licence that is completed by the hunter after the hunting season and returned to government.

Lignin: the natural glue that holds wood fibres together.

Limiting factor: an element of the environment that limits the size of a population.

Limnologist: ecologist who studies freshwater ecosystems.

Lithosphere: the portion of the biosphere containing rocks and soil.

Longline: a floating line on which crops of mussels grow.

Low-grade ore: ore with a low metal concentration.

Mantle: a semi-solid and shifting layer above the earth's molten inner core.

Mariculture: the use of the marine or saltwater environment, as opposed to freshwater, for the cultivation of aquatic plants or animals.

Marine: associated with the oceans.

Marsh: a nutrient-rich non-wooded wetland.

Meltdown: the result of the overheating of a nuclear power plant's reactor core.

Metal anomaly: an unusually high concentration of metal in rock or soil.

Metal mobilization: the dissolving of metals into solution; e.g. as a result of exposure to acid precipitation.

Mineral: a naturally occurring inorganic compound that has formed from physical and chemical actions and reactions within the earth.

Mineral soil: soil composed primarily of minerals, in contrast to organic soil.

Mineralization: the transformation of a metal into an ore; to convert into mineral or organic form.

Mitigation/mitigative measures: actions taken to reduce or minimize the harmful impacts of human activities or developments.

Monitoring well: a special well around a dump site in which the water is regularly tested.

Mortality: deaths by natural and other causes.

Mortality factors: the causes of death.

NASA: the National Aeronautics and Space Administration, which is responsible for many American space flights.

NATO: North American Treaty Organization.

Natural mortality: natural causes of death such as disease, predation, starvation, old age, injury and severe weather.

NIMBY: Not In My Back Yard.

Non-commercial: for personal use; not for sale.

Nonrenewable resources: nonliving resources such as minerals that cannot renew themselves.

Ocean pollutants: substances that enter the sea which lead to a harmful change such as dirtiness, impurity, unhealthiness or hazard.

Offshore fishery: a fishery conducted from large trawlers that remain at sea for weeks at a time.

Omnivore: animals that eat both plants and other animals.

Open ecosystem: an ecosystem such as a river or lake which is open to changing conditions from the outside.

Organic: material containing carbon as a result of once being alive.

Organic fertilizer: fertilizer made from naturally occurring substances.

Organic soil: soil which contains carbon because it was once part of living tissue (see mineral soil).

Otolith: earbone of a codfish.

Outcrop: bedrock that pokes out above the covering of soil, water or glacial deposits.

Ozone: a bluish gas consisting of molecules that are made up of three oxygen atoms.

Paralytic shellfish poisoning: a type of poisoning that results from eating shellfish that contain the spores of dynoflagellates.

Particulate: tiny solid or liquid particles floating in the air.

Passive solar heating: heating that occurs when the flow of energy is by natural means, such as conduction, convection and radiation.

Peat: the accumulated remains of partially decomposed plants which forms the basis for bogs and fens.

Peatlands: poorly drained areas of land that contain a thick layer of peat; e.g. bogs and fens.

Permafrost: a permanently frozen layer of ground.

Person-year: one person working for one year.

Pesticide: a sustance for destroying pests, especially insects.

pH: the acidity or basicity of a substance.

Phosphates: phosphoric acids used as fertilizer.

Photodegradable: capable of being broken down through exposure to light.

Photosynthesis: the process by which plants convert light energy to chemical energy.

Photovoltaic cells: cells used to produce electricity directly from sunlight.

Phytoplankton: plant plankton.

Pitcher plant: a carnivorous plant adapted to live in nutrient poor areas such as peatlands.

Placer deposit: a place where heavy metals or other minerals have settled out of running water as it slows.

Plankton: organisms floating in water habitats, including algae, bacteria, fish larvae, small crustaceans and other tiny organisms.

Pollutant: an impurity; a contaminant.

Population: a group of the same species living close enough together to allow them to breed with one another.

Population density: the number of organisms per unit area.

Precommercial thinning: removal of trees from a 10- to 15-year-old forest so the best trees are separated by about 2 metres.

Predicted hunter success: percentage of hunters that are likely to successfully harvest an animal.

Primary consumers: animals that live mostly on green plants–herbivores.

Producer: an organism which produces its own organic compounds, e.g. plants.

Productive forest: forested land that is considered to be economically valuable.

Productivity: number of animals added to a population each year.

Proponent: an organization or person wishing to undertake a development.

Public consultation: a process in which citizens are consulted for their points of view.

Quota: a set amount of a resource, such as fish or wildlife, that is allocated for harvesting by a particular group of resource users or within a particular area.

Relative abundance: an estimate of the number of animals or plants in an area, in contrast with absolute abundance.

Renewable resources: living resources such as trees and wildlife that can renew themselves when conditions permit.

Respiration: the process by which plants and animals release chemical energy to do work.

Salmonid: a member of the salmon family, which includes salmon, trout and char.

Salvage cutting: the cutting of large numbers of dead or dying trees after an insect infestation.

Sanitary landfill: an area where wastes are regularly dumped, spread and compacted in a site where there is a natural depression or the ground is trenched.

Saxitoxin: a poisonous substance produced by some dynoflagellates.

Secondary consumer: animals that eat herbivores.

Sedges: plants similar to grasses but with three-sided stems instead of flat stems like grasses.

Selection cutting: cutting a small portion of the trees in a forest stand during any one year.

Sewage sludge: precipitated solid matter produced by water and sewage treatment processes.

Shallow water wetlands: non-fluvial bodies of standing water (ponds).

Shellfish: marine organisms that have shells, such as mussels, clams and scallops.

Silviculture: the branch of forestry that deals with the development, cultivation and reproduction of trees.

Simple landfill: a designated area where users may dump their garbage; this garbage is then buried using a bulldozer.

Slash-and-burn farming: the practice of clearing a small plot of land and burning the trees and limbs, the ashes of which enrich the soil.

Slurry: ore that is finely ground and mixed with water.

Sphagnum moss: a sponge-like moss, found in peatlands, which provides the foundation or platform upon which all other plants grow.

Spruce budworm: the caterpillar of a small brown moth that feeds on the needles of balsam fir and certain species of spruce.

Steelhead trout: a kind of rainbow trout that spends part of its life at sea.

Sterile: unable to breed.

Strip and block cutting: cutting by narrow channels or blocks within a forest.

Succession: the gradual change from one plant community to another.

Sulphur dioxide: a gas that occurs naturally in the atmosphere, but is elevated to unhealthy levels by certain human activities.

Sundew: a carnivorous plant adapted to live in nutrient poor areas such as peatlands.

Sustainable development: the development of our resources to meet our present needs without reducing the ability of future generations to meet theirs.

Swamp: a nutrient-rich wooded wetland.

Sweep count: a method of measuring the number of animals in a given area.

Tailings: waste products resulting when activated carbon is extracted from slurry.

Terrestrial: associated with the land.

Tertiary consumer: animals that eat other carnivores.

Thermal pollution: a harmful discharge of a warmer or colder substance, usually water, which is significant enough to raise or lower the temperature of a river, lake or ocean.

Thermomechanical processing: the conversion of wood chips to pulp through a cooking and grinding process.

Thinning: cutting certain trees within a forest stand to reduce competition among the remaining trees.

Timber management: the managing of forests primarily for wood products.

Total allowable catch: the total amount of a particular fish species allowed to be caught from a particular stock for a specific period of time, as measured in metric tonnes.

Total suspended solid: a measure of the fine waste particles in a quantity of water.

Toxic: poisonous, can cause illness or death to plants and animals.

Transboundary pollution: pollution that is not contained within the country where it originates.

Trend data: often used to measure the trend of a population; whether it is growing, declining or staying the same.

Triploid: having three chromosomes.

Trophic level: position in a food or energy chain determined by the number of supporting levels.

Tundra: a treeless arctic area usually underlain by permafrost; may be wet or dry.

Ultraviolet radiation: part of the natural radiation from sunlight; the wavelengths have slightly higher energy and are slightly shorter than visible light.

Understory: the plants of a forest growing beneath the main canopy.

Unsustainable development: the over-development of our resources, reducing the ability of future generations to meet their needs.

Virtual population analysis: a method used for estimating population size of a species by looking back over time at the numbers of fish caught of a particular year-class, and adding to this an estimate of natural mortality.

Watershed: the entire drainage area of a river and its tributaries.

Weed species: plants that compete with or harm the type of trees that are desirable in a forest stand.

Wetlands: areas of land saturated with water including bogs, fens, swamps, marshes, and shallow water ponds.

Year-class: animals that hatch or are born in the same year and are therefore the same age.

Yield: the quantity of biomass produced in a specified time.

Zooplankton: animal plankton.

Index

Illustration Credits

For each entry below, page numbers are followed by a colon, the source name (or names, separated by a semi-colon) and a slash to indicate end of entry. Where more than one illustration appears on a page, the credit order is from the top to the bottom of the page. The following abbreviations appear for names that occur frequently:

DM - Dennis Minty
SJM - Susan J. Meades
NWF - National Wildlife Federation
WLD - Wildlife Division, Dept of Tourism and Culture
WJM - William J. Meades
DFO - Department of Fisheries and Oceans

15: Dan Murphy/ 17-18: Sylvia Ficken/ 19: DM/ 20: NASA; Sylvia Ficken/ 21: Save the Birds/ 22-23: Sylvia Ficken/ 24: NASA/ 25: SJM/ 27: Space Biospheres Ventures/ 28: The Biodome of Montreal/ 29: SJM (map [adapted from H. Walter, 1973. *Vegetation of the Earth*, Springer-Verlag, and J.S. Rowe, 1972. *Forest Regions of Canada*, Information Canada, Ottawa] and forest); WJM (tundra)NWF (desert and grassland); DM (ocean, freshwater)/ 30: SJM/ 32: NWF/ 33: SJM/ 34: DM/ 35: Save the Birds/ 38-43: SJM/ 47: Barry May/ DM/ 48: DM/ 49: DM; SJM/ 50-51: SJM/ 52: DM; SJM/ 53: SJM/ 54: DM; ZESA, with permission from Stoddart Publishing/ 55: SJM/ 56: SJM; DM/ 57: Bernard Ball/ 58-61: DM/ 62-64: SJM/ 65: DM; SJM/ 66: SJM/ 67: DM/ 68: WLD/ 69: Breakwater/ 70-71: WLD/ 72: Terra Nova National Park/ 76: WJM (1,2); SJM (3,4)/ 77: SJM (1,3,4,5); DM (2)/ 78-79: DM (1); SJM (2,6); WJM (3-5, 7-10)/ 84-95: DFO/96: Michel Therien/ 97: DM/ 98: Dept of Environment and Lands/ 101: DM/ 102: Environment and Lands/ 110: The Royal Netherlands Air Force/ 113: Anikashan Antane/ 114: CFB Goose Bay/ 123-130: WLD/ 138-146: DM/ 152: NWF/ 157: DM; Dan Murphy/ 162: Richard Wheeler/ 163: DM/ 166: Dan Murphy/ 167: SJM/ 170: Rick West, Forestry Canada/ 176: CBPP/ 181: Forestry Canada/ 190-198: Baxter Kean, Dept of Mines and Energy/ 201: W.Skinner/ 203: DFO/ 219-225: Jim Carscadden/ 230: DM/ 234: DFO, Burlington/ 242: Industry, Science and Technology Canada/ 254: Atmospheric Environment Service, Environment Canada/ 265: R.D. Elliot/ 267: Zoe Lucas/ 277: Nfld. Hydro/ 283: DM/ 290-292: DM.

Figure Credits

All figures/tables not credited below are the work of the authors, based on data from a wide variety of sources.

Figs. 2.1 - 2.4: Susan J. Meades; Table 2.1: Adapted from *Wetlands of Canada*, Environment Canada, 1988; Fig. 3.1: Adapted from DFO Calendar, 1992; Fig. 3.2: Adapted from "The Science of Cod," DFO, 1988; Fig. 5.1: From Goose Bay EIS; Fig. 7.2: Parks Division, Dept of Tourism and Culture; Figs. 8.1 - 8.2: Wildlife Division, Dept of Tourism and Culture; Fig. 9.2, 9.4: From *The Newfoundland and Labrador Environment - A Collection of Case Studies*, used by permission of the publisher Jesperson Press Limited; Table 10.1, Fig. 10.3: Data from Industrial Environmental Engineering Division, Department of Environment and Lands, Report on Corner Brook Airborne Particulate 1990-1991; Figs. 12.2-12.3: From "A Guide to Longline Mussel Culture in Newfoundland," DFO, 1988; Table 12.2: Data from Newfoundland and Labrador Fisheries Statistics, Courtesy of Aquaculture Division; Fig. 13.1: NWF; Figs. 14.1, 15.2, 15.3: From Government of Canada, 1991, *The State of Canada's Environment*. Reproduced with permission of the Minister of Supply and Services Canada, 1993.

List of Technical Reviewers

The following people generously assisted in the development of this text by providing technical reviews of a draft edition of specific chapters. We are indebted to them all for their valuable comments and criticisms. The authors have made all efforts to accommodate and balance matters of opinion, yet make no claims to have pleased all concerned.

SUSAN KAINS AHEARN, Faculty of Education: Science Education, MUN; WILLIAM J. MEADES, Program Director, Forest Resources & Environmental Research, Forestry Canada, St. John's; JON LIEN, Whale Research Group, MUN; BILL MONTEVECCHI, Department of Psychology, Animal Behaviour, MUN; KEN DOMINIE, Director, Civil Sanitary Environmental Engineering, Provincial Dept of Environment and Lands; LIEUTENANT L.R. PITTMAN, Public Affairs Officer, Department of National Defence, CFB Goose Bay; DANIEL ASHINI, Director of Innu Rights and Environment, Sheshatshiu, Labrador; SHANE MAHONEY, Wildlife Division, Dept of Environment and Lands; DONALD G. HUSTINS, Director, Parks Division, Department of Tourism and Culture; OSCAR FORSEY, Wildlife Division, Dept of Environment and Lands; JIM TAYLOR, Section Head, Forest Management Planning, Department of Forestry and Agriculture, Corner Brook; R.J. WEST and WADE BOWERS, Research Scientists, Forestry Canada, St. John's; GEORGE VANDUSEN, Chief Forester, Corner Brook Pulp and Paper Ltd.; FERD MORRISSEY, Manager, Engineering, Dept of Mines and Energy; SCOTT SWINDEN, Dept of Mines and Energy; KEN ANDREWS, Director, Mineral Lands Division, Dept of Mines and Energy; PAUL DEAN, ADM, Mineral Resource Management Branch, Dept of Mines and Energy; COLIN B. McKENZIE, Mineral Consultant, Corner Brook; PETER DIMMELL, Exploration Consultant, St. John's; TOM LANE, Teck Explorations, St. John's; LARRY CONNELL, Manager of Environmental & Metallurgical Services, Royal Oak Mines Inc., Vancouver; REX PORTER, Division Head, Salmonid and Habitat Science, DFO, St. John's; TIM BOYLE, Co-ordinator, Genetics and Biodiversity, Forestry Canada, Ottawa; PETER J. SCOTT, Associate Professor and Curator, Ayre Herbarium, MUN; M.G. WEBER, Research Scientist, Forestry Canada, St. John's; ROBERT P. WHELAN, Environmental Laboratory Chemist, Industrial Environmental Engineering Division, Dept of Environment and Lands; BILLIE L. BEATTIE, Meteorologist, Atmospheric Environment Service, Environment Canada, Bedford, NS; JOHN D. JACOBS, Professor and Department Head of Geography, MUN; JOHN DUBLIN, Meteorologist, Atmospheric Environment Service, Environment Canada, Bedford, NS; DERRICK MADDOCKS, Director, Industrial Environmental Engineering, Dept of Environment and Lands; JOHN W. CHARDINE, Research Scientist, Canadian Wildlife Service, St. John's; KEVIN C. POWER, Head, Environmental Engineering, Environment Canada, St. John's; JERRY F. PAYNE, Head, Toxicology and Contaminants, Science Branch, DFO, St. John's; JOHN T. DROVER, Manager, Energy Demonstration and Development, Dept of Mines and Energy; NICK MARTY, Policy Development and Analysis, Energy, Mines and Resources Canada; ANGUS A. BRUNEAU, Chairman and CEO, Fortis Inc.; Chairman, Newfoundland Power.

303